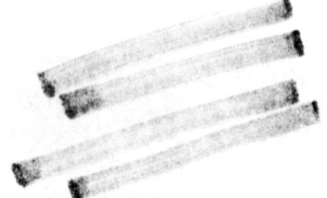

The following States are Members of the International Atomic Energy Agency:

AFGHANISTAN	HOLY SEE	PANAMA
ALBANIA	HUNGARY	PARAGUAY
ALGERIA	ICELAND	PERU
ARGENTINA	INDIA	PHILIPPINES
AUSTRALIA	INDONESIA	POLAND
AUSTRIA	IRAN, ISLAMIC REPUBLIC OF	PORTUGAL
BANGLADESH	IRAQ	QATAR
BELARUS	IRELAND	ROMANIA
BELGIUM	ISRAEL	RUSSIAN FEDERATION
BOLIVIA	ITALY	SAUDI ARABIA
BRAZIL	JAMAICA	SENEGAL
BULGARIA	JAPAN	SIERRA LEONE
CAMBODIA	JORDAN	SINGAPORE
CAMEROON	KENYA	SLOVENIA
CANADA	KOREA, REPUBLIC OF	SOUTH AFRICA
CHILE	KUWAIT	SPAIN
CHINA	LEBANON	SRI LANKA
COLOMBIA	LIBERIA	SUDAN
COSTA RICA	LIBYAN ARAB JAMAHIRIYA	SWEDEN
COTE D'IVOIRE	LIECHTENSTEIN	SWITZERLAND
CUBA	LUXEMBOURG	SYRIAN ARAB REPUBLIC
CYPRUS	MADAGASCAR	THAILAND
CZECHOSLOVAKIA	MALAYSIA	TUNISIA
DEMOCRATIC PEOPLE'S	MALI	TURKEY
REPUBLIC OF KOREA	MAURITIUS	UGANDA
DENMARK	MEXICO	UKRAINE
DOMINICAN REPUBLIC	MONACO	UNITED ARAB EMIRATES
ECUADOR	MONGOLIA	UNITED KINGDOM OF GREAT
EGYPT	MOROCCO	BRITAIN AND NORTHERN
EL SALVADOR	MYANMAR	IRELAND
ESTONIA	NAMIBIA	UNITED REPUBLIC OF
ETHIOPIA	NETHERLANDS	TANZANIA
FINLAND	NEW ZEALAND	UNITED STATES OF AMERICA
FRANCE	NICARAGUA	URUGUAY
GABON	NIGER	VENEZUELA
GERMANY	NIGERIA	VIET NAM
GHANA	NORWAY	YUGOSLAVIA
GREECE	PAKISTAN	ZAIRE
GUATEMALA		ZAMBIA
HAITI		ZIMBABWE

The Agency's Statute was approved on 23 October 1956 by the Conference on the Statute of the IAEA held at United Nations Headquarters, New York; it entered into force on 29 July 1957. The Headquarters of the Agency are situated in Vienna. Its principal objective is "to accelerate and enlarge the contribution of atomic energy to peace, health and prosperity throughout the world".

© IAEA, 1993

Printed by the IAEA in Austria
January 1993

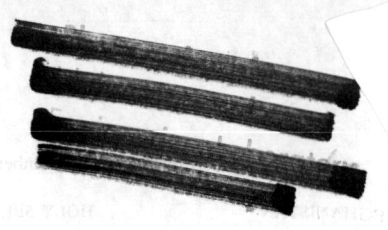

IMPROVED CEMENT SOLIDIFICATION
OF LOW AND INTERMEDIATE LEVEL
RADIOACTIVE WASTES

TECHNICAL REPORTS SERIES No. 350

IMPROVED CEMENT SOLIDIFICATION OF LOW AND INTERMEDIATE LEVEL RADIOACTIVE WASTES

INTERNATIONAL ATOMIC ENERGY AGENCY
VIENNA, 1993

VIC Library Cataloguing in Publication Data

Improved cement solidification of low and intermediate level radioactive
wastes. — Vienna : International Atomic Energy Agency, 1993.
    p. ; 24 cm. — (Technical reports series, ISSN 0074-1914 ; 350)
STI/DOC/10/350
ISBN 92-0-100493-1
Includes bibliographical references.

    1. Radioactive waste disposal.   2. Cementation (Petrology).   I. Inter-
national Atomic Energy Agency.   II. Series: Technical reports series
(International Atomic Energy Agency) ; 350.

VICL   92-00051

FOREWORD

Cementation was the first and is still the most widely applied technique for the conditioning of low and intermediate level radioactive wastes. Compared with other solidification techniques, cementation is relatively simple and inexpensive. However, the quality of the final cemented waste forms depends very much on the composition of the waste and the type of cement used. Different kinds of cement are used for different kinds of waste and the compatibility of a specific waste with a specific cement type should always be carefully evaluated.

Cementation technology is continuously being developed in order to improve the characteristics of cemented waste in accordance with the increasing requirements for quality of the final solidified waste. Various kinds of additives and chemicals are used to improve the cemented waste forms in order to meet all safety requirements.

This report is meant mainly for engineers and designers, to provide an explanation of the chemistry of cementation systems and to facilitate the choice of solidification agents and processing equipment. It reviews recent developments in cementation technology for improving the quality of cemented waste forms and provides a brief description of the various cement solidification processes in use.

The original draft of the report was prepared by consultants P. Colombo (United States of America) and R. Köster (Germany) and then discussed by experts from twelve Member States and two international organizations (the Commission of the European Communities and the International Union of Producers and Distributors of Electrical Energy) at a Technical Committee meeting held in May 1987. The final version of the report was prepared by the IAEA Secretariat and P. Colombo as the IAEA consultant on the basis of comments received from Member States.

The IAEA wishes to express its gratitude to all those who took part in the preparation of this report and is particularly grateful to P. Colombo, who made a valuable contribution at all stages of the preparation of the report.

The responsible officer in the IAEA was V.M. Efremenkov, from the Division of Nuclear Fuel Cycle and Waste Management, with contributions at later stages from A.F. Tsarenko and V.S. Tsyplenkov, also from the Division of Nuclear Fuel Cycle and Waste Management.

EDITORIAL NOTE

CONTENTS

1. INTRODUCTION

It has long been common practice to incorporate radioactive wastes generated at nuclear power plants and fuel reprocessing facilities into cement. In particular, the use of cement to immobilize concentrates from water treatment and liquid purification has been an attractive procedure because of the strength and durability of properly set compositions. Volume reduction processes are often employed before solidification to reduce the quantities of waste requiring immobilization and disposal. Although other solidification agents or binders, such as bitumen and polymers, have gained favour over the years, cement has been the single most commonly used material for conditioning low and intermediate level radioactive wastes. The following are some of the advantages and disadvantages of cement for the solidification of radioactive wastes.

Advantages

- Material and technology well known;
- Compatible with many types of waste;
- Most aqueous wastes chemically bound to matrix;
- Low cost of cement;
- Good self-shielding;
- No vapour problems;
- Long shelf life of cement powder;
- Good impact and compressive strengths;
- Low leachability for some radionuclides;
- No free water if properly formulated;
- Rapid, controllable setting, without settling or segregation during curing.

Disadvantages

- Some wastes affect setting or otherwise produce poor waste forms.
- pH adjustment of waste may be necessary.
- Swelling and cracking occur with some products when they are exposed to water.
- Volume increase and high density may develop.
- Excessive heat may develop during setting with certain combinations of cement and waste.
- Dust problems may occur with some systems.
- Equipment for powder feeding is difficult to maintain.
- Potential maintenance problems may result from premature cement setting, especially in the case of in-line mixers.

The chemical and physical properties of cement are well known and experience in using it for construction extends over many decades. However, knowledge of the effect of mixing cement with certain radioactive wastes is limited and experimental work is often necessary to optimize cement–waste formulations. The situation becomes more complicated when cement is used to solidify nuclear power plant wastes, owing to their widely variable chemistry. For example, boric acid waste retards the setting of Portland cement and, if sufficient quantities are present, will prevent setting. Solidification of detergent concentrates and organic liquids with cement can present difficulties but these may generally be avoided by mixing different waste streams and diluting the problem constituents.

Nevertheless, cement technology has been developed to meet the demands for improved quality of the solidified waste forms. The introduction of special types of cements and admixtures and advanced pretreatment techniques and the growing understanding of the chemical interactions between waste and cement have renewed interest in the applicability of improved cement solidification compositions and processes to specific problem wastes and disposal options.

This report describes current technologies and processes for the improved solidification of low and intermediate level radioactive wastes with cement. Recent developments which enhance the volumetric efficiency and improve the waste form properties by using various types of cements and admixtures are emphasized. In addition, the report describes in some detail the chemical and physical nature of cement, and the effect of various types of waste and waste components on its setting and curing. The specific composition of many liquid waste streams generated at different nuclear facilities and the effect that each of the wastes may have on cement are beyond the scope of this report.

A glossary of the terminology used in cement chemistry is provided at the end of this book.

2. WASTE CHARACTERIZATION

2.1. INTRODUCTION

Earlier publications in the IAEA Technical Reports Series [1–5] have described in great detail the types, sources and characteristics of low and intermediate level radioactive wastes arising from nuclear fuel cycle operations. This section describes the physical and chemical characteristics of the major liquid and wet solid waste streams generated at nuclear power plants.

Wet solid wastes can be classified into four basic types: spent ion exchange resins, precoat filter sludges, evaporator concentrates and miscellaneous liquids.

In the collection of wet solid wastes, various waste streams may be combined in a single tank, or they may be completely segregated. Liquid wastes, including regenerant solutions, boric acid solutions, floor drain waste, laboratory waste and decontamination solutions, are often combined during collection. Demineralizer resins and filter sludges are routinely collected in separate tanks.

Compactible and non-compactible wastes, the major contributor to the annual waste arisings of a plant, are not discussed in this section because they are not process waste. A complete discussion of compactible and non-compactible wastes, or trash, can be found in Refs [1–9].

2.2. SOURCES AND TYPES OF WASTE

2.2.1. Ion exchange resins

2.2.1.1. Physical and chemical properties

Ion exchange resins are porous beads of polystyrene cross-linked with divinylbenzene. Strong cation resins generally contain bound sulphonic acid functional groups and, to a lesser degree, carboxylic or phosphoric acid. Strong anion resins generally contain bound quaternary ammonium functional groups.

A hydrogen ion (H^+) is generally the ionic form of cation resins, although lithium is also found. The ionic form of anion resins is usually a hydroxide (OH^-) form, although chloride (Cl^-) and carbonate (CO_3^{2-}) are also used. The total ion exchange capacity of a resin is defined in terms of equivalent kilograms of calcium carbonate ($CaCO_3$) per cubic metre of resin. Cation resins, with bead size ranging from 0.45 to 0.60 mm, are rated at 1100–1400 kg/m^3. Anion resins, with sizes of 0.38–0.45 mm, are rated at 530–1100 kg/m^3. Mixed bed resins may have exchange capacities as low as 350 kg/m^3.

Anion resins can generally withstand temperatures as high as 93–121°C. If oxygen is present in the water, the temperature limit may be as low as 38°C, depending on the type of resin.

When shipped, the resins are dry (without free water) and fully swollen. The actual material contains between 42 and 55% water by weight.

2.2.1.2. Applications in LWRs

The major applications of deep bed resins in LWRs are listed below [5]:

BWRs

— Liquid radioactive waste treatment
— Condensate polishing
— Spent fuel pool cleanup
— Chemical laboratory.

3

PWRs

  — Liquid radioactive waste treatment
  — Condensate polishing
  — Spent fuel pool cleanup
  — Chemical laboratory
  — Chemical and volume control
  — Boron control.

From the standpoint of waste volume generation, the most significant application of deep bed resins is in the radioactive waste treatment system and in the condensate polishing system. When exhausted, waste treatment ion exchange resins are not regenerated. When condensate polishing resins are exhausted, they are regenerated and reused. The chemicals from regeneration are sent to the liquid radioactive waste treatment system for processing.

2.2.2. Precoat filter sludges

2.2.2.1. Precoat filter materials

Sludge from precoat filters is a combination of the original precoat material, the insolubles removed from the inflowing stream such as dirt, corrosion particles and other suspended solids, and flocculating agents (filter aids) used to extend the life of the filters. Numerous types of precoat material are available. The most commonly used are Solka-Floc, a cellulose fibre derived from wood pulp; diatomaceous earth or diatomaceous silica, the siliceous skeletons of microscopic aquatic plants; and Powdex, a ground ion exchange resin manufactured specifically for use on precoat filters.

2.2.2.2. Applications in LWRs

The major applications of precoat filters in LWRs are listed below:

BWRs

  — Condensate polishing
  — Liquid radioactive waste treatment
  — Reactor water cleanup
  — Spent fuel pool cleanup.

PWRs

  — Condensate polishing
  — Spent fuel pool cleanup.

Precoat filters are used primarily in BWRs because of their larger capacity and higher crud holding capacity compared with cartridge filters.

In BWRs the majority of the precoat filter sludge volume is from the radioactive waste treatment system filters. The waste from the reactor water cleanup system contributes the largest share of the activity. Powdex is used in almost all reactor water cleanup systems. This accounts for the presence of soluble fission products such as caesium. Most spent fuel pool cleanup systems use precoat filters, many with Powdex, which may account for the relatively high concentration of transuranics in precoat filter waste.

The precoat filters used for condensate polishing are the largest of all the precoat filters used in LWRs. As such, they also have the largest quantity of precoat and dirt at the time of backwash. However, unless there is significant inleakage of water with a high undissolved content to the condenser, these filters are backwashed only about once every month or two. On the other hand, backwashing of the reactor water cleanup system is carried out twice a week.

2.2.3. Evaporator concentrates

2.2.3.1. Boric acid concentrates

Evaporators used in most PWRs are not capable of concentrating boric acid solutions beyond 12.5% by weight. At higher concentrations the boric acid begins to crystallize, a condition for which these evaporators were not designed. Newer evaporators, aptly named crystallizers, are designed to handle evaporator bottoms of up to 50% boric acid by weight. The pH of the evaporator bottoms averages 6.5 and ranges from 4 to 9. The density of this waste is 1.0 g/cm^3, increasing to approximately 1.6 g/cm^3 when the waste is solidified.

2.2.3.2. Sodium sulphate concentrates

Virtually all evaporator bottoms containing sodium sulphate come from BWRs. Current generation radioactive waste evaporators are limited to producing bottoms at 25% sodium sulphate by weight. As in the case of boric acid solutions, crystallizers will concentrate these wastes up to 50% sodium sulphate by weight. Other materials in the evaporator bottoms include defoaming agents, cleaning solutions and possibly some excess sodium hydroxide from the regeneration of deep bed demineralizer resins. Sodium sulphate wastes range in pH from 4.5 to 9 or above, and range in density from 1.2 g/cm^3 unsolidified to 1.45 g/cm^3 when solidified.

2.2.3.3. Miscellaneous chemical concentrates

These wastes are found in both PWRs and BWRs and are the result of the concentration of chemical wastes such as decontamination solutions, chemical laboratory drain waste, dilute water chemistry chemicals such as those used to adjust the primary coolant system pH, and laundry waste water. When concentrated, these wastes are not limited by the crystallization of boric acid and can be concentrated up to 25% solids by weight. The presence of laundry soap in the waste may account for a low weight per cent solids. The pH of these wastes is between 8 and 10, although it may be less than 6.5, probably because of boron in the form of boric acid.

2.2.3.4. Applications of evaporators in LWRs

Evaporators in BWRs are found only in the radioactive waste treatment system, where they are used to concentrate sodium sulphate (resulting from the regeneration of condensate demineralizers), decontamination solutions, laundry waste water and floor drain waste. Regeneration wastes and decontamination solutions may be collected in separate tanks to allow for independent pretreatment, such as pH adjustment, but they are usually processed through the same evaporator. The bottoms are then collected in the same tank, making identification of the operating characteristics of the evaporators (relative to the waste processed) almost impossible.

In PWRs, evaporators are used to concentrate boron in the boron recovery system for storage and reuse, and to remove impurities from the steam generator blowdown. The condensed vapour from the steam generator blowdown may be either continuously released or recycled to the plant. Evaporators are also used in the liquid and chemical radioactive waste treatment system in much the same way as is the laundry waste evaporator in BWRs. The systems in which evaporators are used in LWRs are listed below:

BWRs

— Chemical waste treatment
— Floor drain waste treatment
— Laundry waste treatment.

PWRs

— Boron recovery
— Steam generator blowdown
— Liquid waste treatment
— Chemical waste treatment
— Laundry waste treatment.

3. CHARACTERIZATION OF CEMENT FOR SOLIDIFICATION OF WASTE

3.1. INTRODUCTION

There is a wide range in the composition of cementitious materials used as solidification agents for radioactive wastes. Their common feature is that the main constituents are lime (CaO), silica (SiO_2) and alumina (Al_2O_3).

In Fig. 1 the ternary phase diagram shows the composition limits for Portland cement in the system $CaO-Al_2O_3-SiO_2$. The composition of most modern cements falls within the triangle $C_3S-C_2S-C_3A$ in the Portland cement zone (see Table I).

Portland cements, which in most countries are manufactured with materials that are locally available, are the most commonly used type of cement. Other readily available types of cement considered in this section are Portland blast-furnace cement, high alumina cement and masonry (high lime) cement.

3.2. TYPES OF CEMENT AND THEIR PROPERTIES

3.2.1. Ordinary Portland cement

Ordinary Portland cement is composed chiefly of three oxides: silica (SiO_2), lime (CaO) and alumina (Al_2O_3), with small quantities of magnesia (MgO), ferric oxide (Fe_2O_3), sulphur trioxide (SO_3) and other oxides introduced as impurities in the raw materials used in its manufacture. Portland cements are produced by heating ground calcareous (lime) and argillaceous (clay) materials at a clinkering temperature (1400–1600°C) and grinding the resulting clinker. Calcareous materials are obtained from limestone, sea shells and marls, and argillaceous materials are obtained from clay, shale, bauxite, silica sand and iron ore. Clay and shale are usually added when alumina is not present in limestone in sufficient amounts. The cements are not simply mixtures of these oxides but are mixtures of combinations of the basic oxides [10, 11].

3.2.1.1. Composition

The four basic compounds in Portland cement are tricalcium silicate ($3CaO \cdot SiO_2$), dicalcium silicate ($2CaO \cdot SiO_2$), tricalcium aluminate ($3CaO \cdot Al_2O_3$) and tetracalcium aluminoferrite ($4CaO \cdot Al_2O_3 \cdot Fe_2O_3$). Table I gives the oxide composition and the abbreviations universally used by chemists working in this field.

7

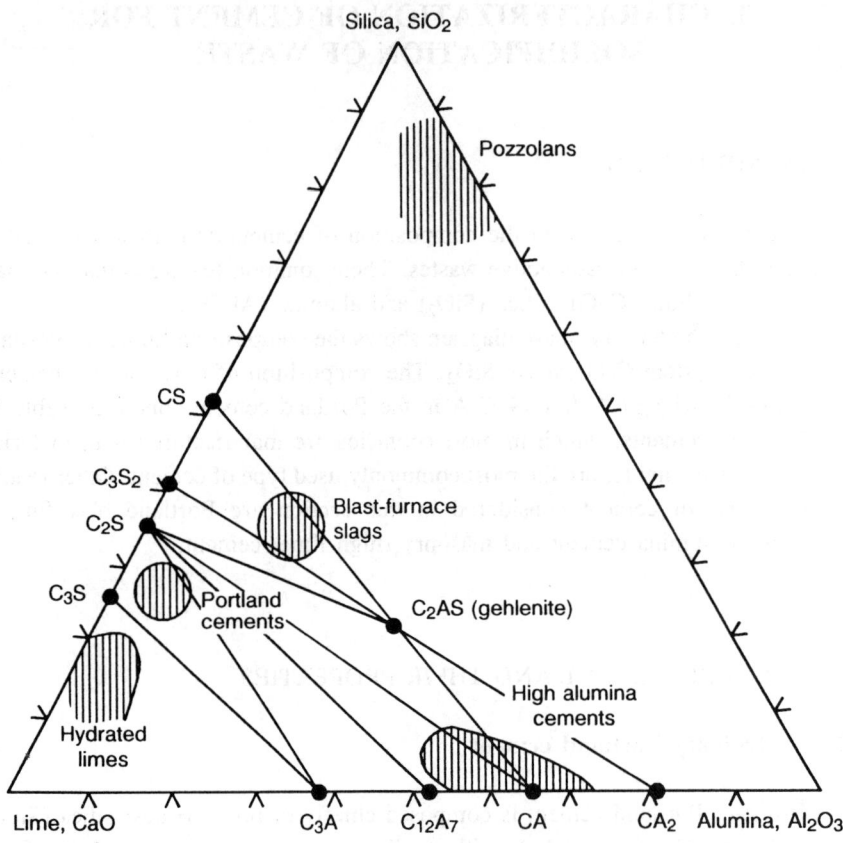

FIG. 1. Cement compositions [10].

TABLE I. PRINCIPAL COMPOUNDS IN PORTLAND CEMENT [11, 12]

Compound	Oxide composition	Abbreviation
Tricalcium silicate	$3CaO \cdot SiO_2$	C_3S
Dicalcium silicate	$2CaO \cdot SiO_2$	C_2S
Tricalcium aluminate	$3CaO \cdot Al_2O_3$	C_3A
Tetracalcium aluminoferrite	$4CaO \cdot Al_2O_3 \cdot Fe_2O_3$	C_4AF

TABLE II. COMPOSITION (%) OF PORTLAND CEMENTS [13]

Type of cement	C_3S	C_2S	C_3A	C_4AF
I Normal	45	27	11	8
II Modified	44	31	7	13
III High early strength	53	19	10	7
IV Low heat	20	52	6	14
V Sulphate resistant	38	43	4	8

Each of the compounds in Table I contributes to the behaviour of cement. By varying the concentration of each component, a cement having special properties can be made. Since a large number of cements can be made by varying compositions, national standards specify several types. The compositions of five types of Portland cement, as specified in the United States of America by the American Society for Testing and Materials (ASTM) [13], are given in Table II.

Type I, normal Portland cement, is the most commonly used, because of its high availability and low cost. It is generally used where the special properties of the other types are not required, where it will not be subject to sulphate attack from soil or water, or where the heat generated by the hydration of the cement will not cause an unacceptable rise in temperature.

Type II, modified Portland cement, has a lower heat of hydration than Type I and generates heat at a slower rate. It also has improved resistance to sulphate attack and is used where added precautions are needed, as in groundwaters where sulphate concentrations are higher than normal.

Type III, high early strength cement, develops strength rapidly as a result of its high tricalcium aluminate and tricalcium silicate contents. However, this development is accompanied by a high rate of heat evolution, which may preclude use of this type for massive waste–cement monoliths.

Type IV cement is a low hydration heat cement that is used primarily for massive waste–cement monoliths. The low rate of heat evolution in this cement type is attributable to its high dicalcium silicate content and corresponding low content of tricalcium silicate and tricalcium aluminate.

Type V cement is sulphate resistant as a result of its low tricalcium aluminate content. It is intended for use in monoliths exposed to severe sulphate action, such as in soils and waters of high alkaline content. It gains strength at a slower rate than normal Portland cement.

3.2.1.2. Hydration

When Portland cement is mixed with water, its constituent compounds undergo a series of chemical reactions, termed hydration, which eventually causes it to harden.

The chemical reactions occurring during the setting, hydration and ageing of a Portland cement–water mixture are complex, interrelated and not thoroughly understood. As shown below, the two calcium silicates, which constitute about 75% of a Portland cement by weight, react with water to produce two new compounds: calcium hydroxide and a calcium silicate hydrate called tobermorite gel.

$$2(3CaO \cdot SiO_2) \quad + \quad 6H_2O \longrightarrow 3CaO \cdot 2SiO_2 \cdot 3H_2O \quad + \quad 3Ca(OH)_2$$

(tricalcium silicate)   (water)   (tobermorite gel)   (calcium hydroxide)

$$2(2CaO \cdot SiO_2) \quad + \quad 4H_2O \longrightarrow 3CaO \cdot 2SiO_2 \cdot 3H_2O \quad + \quad Ca(OH)_2$$

(dicalcium silicate)   (water)   (tobermorite gel)   (calcium hydroxide)

The tricalcium aluminate and tetracalcium aluminoferrite combine with considerably more water on a molar basis than do the calcium silicate compounds.

$$3CaO \cdot Al_2O_3 \quad + \quad 12H_2O \quad + \quad Ca(OH)_2 \longrightarrow 3CaO \cdot Al_2O_3 \cdot Ca(OH)_2 \cdot 12H_2O$$

(tricalcium aluminate)   (water)   (calcium hydroxide)   (tetracalcium aluminate hydrate)

$$4CaO \cdot Al_2O_3 \cdot Fe_2O_3 \quad + \quad 10H_2O \quad + \quad 2Ca(OH)_2 \longrightarrow 6CaO \cdot Al_2O_3 \cdot Fe_2O_3 \cdot 12H_2O$$

(tetracalcium aluminoferrite)   (water)   (calcium hydroxide)   (calcium aluminoferrite hydrate)

The reaction between tricalcium aluminate, water and gypsum (a compound present in cement) produces a calcium sulphoaluminate.

$$3CaO \cdot Al_2O_3 \quad + \quad 10H_2O + CaSO_4 \cdot 2H_2O \longrightarrow CaSO_4 \cdot 12H_2O \cdot 3CaO \cdot Al_2O_3$$

(tricalcium aluminate)   (water)   (gypsum)   (calcium monosulphoaluminate)

Each of the compounds formed plays an important role in determining the properties of concrete. By far the most important compound is tobermorite gel, which is the main cementing component of concrete, and on which properties such as setting and hardening, strength and dimensional stability primarily depend.

The relative rates of reaction of the individual phases are shown in Fig. 2. The hydraulic value of cement, however, depends on a number of factors such as the chemical composition of the clinker, the sintering temperature, the fineness of grinding, the presence of impurities and the effects of waste composition.

In the setting and hardening process of the cement binder, three phases can be distinguished:

(1) During the absorption of water the hydrated mineral compounds form a colloidal disperse substance called a sol.

(2) Sol coagulates according to the rules of colloidal chemistry into a gel substance and precipitates as soon as its static electric charge is lost. Phases (1) and (2) represent the hydration and setting of cement. The beginning of the setting period corresponds approximately to the start of gelatinization. The end of the setting period coincides with the end of the precipitation of floccules. The entire process, up to the attainment of a friable rigidity, is referred to as setting.

(3) When the two processes are completed, the gel starts to dry and to crystallize in the form of slabs and needles, a process called hardening.

The hydration reactions of Portland cement are all exothermic, resulting in the liberation of heat; the extent of the rise in temperature depends on how quickly heat is liberated from the waste form. The contribution of each compound to the overall rate of heat evolution is a function of the heat of hydration, the rate of hydration and the fraction of the compound in the cement. The hydration characteristics of the cement compounds are summarized in Table III.

The hydration process is complex, involving both hydration and hydrolysis, and often forming metastable hydration products (Fig. 3). A typical plot of the rate of heat evolution measured by conduction calorimetry for a paste of Portland cement in water is given in Fig. 4.

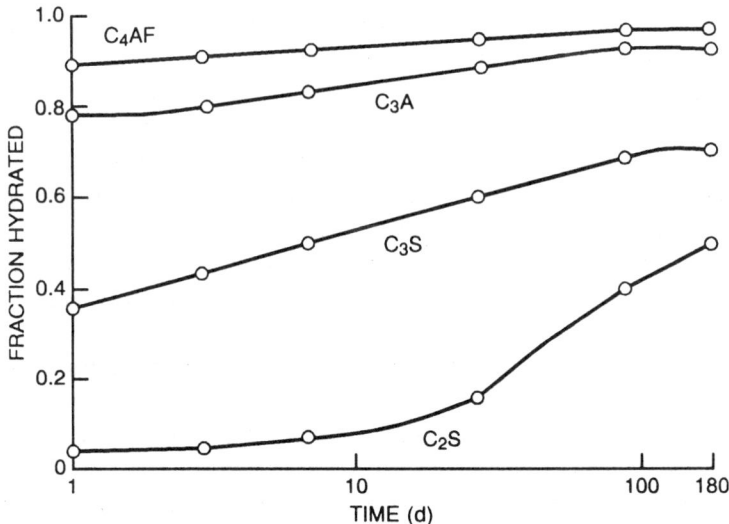

FIG. 2. *Rate of hydration of pure Portland cement compounds [14].*

TABLE III. HYDRATION CHARACTERISTICS OF THE COMPOUNDS OF PORTLAND CEMENT [11, 12]

	C_3S	C_2S	C_3A	C_4AF
Heat of hydration (cal/g)[a]	120	62	207	100
Rate of hydration (fraction on first day)	0.4	0.05	0.8	0.9
Temperature increase	Medium	Low	High	Medium
Rate of compressive strength development	Medium	Slow	Fast	Fast
Early strength (28 d) (MPa)	47	5	3	2
Ultimate strength (360 d) (MPa)	72	71	8	5

[a] 1 cal = 4.186 J.

3.2.1.3. General properties

Porosity

During solidification, the volume of a cement–waste mixture decreases, i.e. the mixture shrinks. As a result of shrinkage due to the drying of the gel substance or to crystallization, gel pores are formed. Capillary pore systems are also formed from the evaporation of excess, unbound water.

The gel pores range in diameter from 10 mm to less than 0.0005 mm. Their volume percentage increases with hydration and attains about 20–30% of the cement paste fraction in hardened cement paste. Capillary pores attain sizes of 1–10 mm, with a volume percentage ranging from 0 to 40%, and increase with the amount of mixing water used in the cement mixture and with the progress of hydration.

The entropies of gel and capillary water are different. Changes in temperature cause a movement of water from the gel into the capillary pores and vice versa, which results in further changes in volume.

In addition to these pores, there are air pores whose size varies from 0.01 to 2 mm. Their volume percentage in hydrated cement may be from 1 to 10%.

The gel pores of cement are insignificant from the point of view of leachability and corrosion. However, the microchannels, which may build up a capillary system, offer a point of attack for corrosion, since they offer passages for the penetration of aggressive water into the interior of the hardened cement. Air pores are not detrimental, if well distributed and sufficiently small. In fact, they may reduce

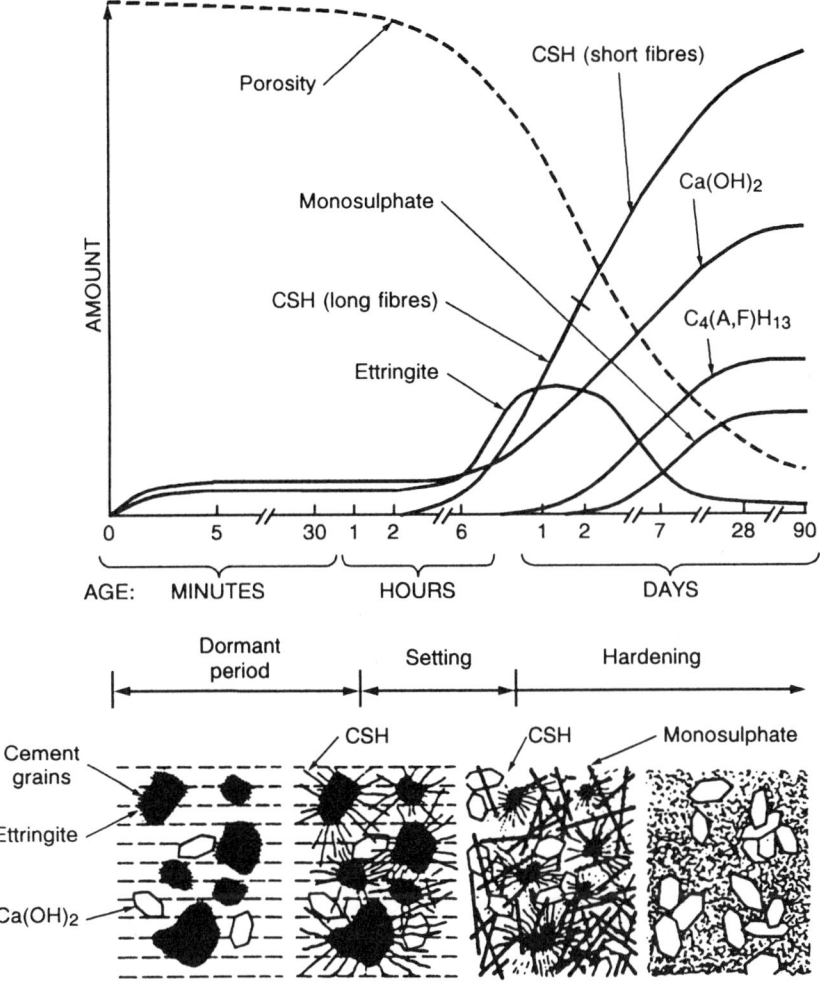

FIG. 3. *Course of cement paste reactions. CSH: calcium silicate hydrate [10].*

permeability by permitting a lower quantity of mixing water to be used. When concrete is submerged in water, the capillary pores fill first. Water can penetrate into the air pores only after air is displaced by diffusion from the capillary pores.

Water/cement ratio

The water/cement (w/c) ratio is probably the most significant single item affecting the strength and chemical resistance of a hardened cement mix. In the

13

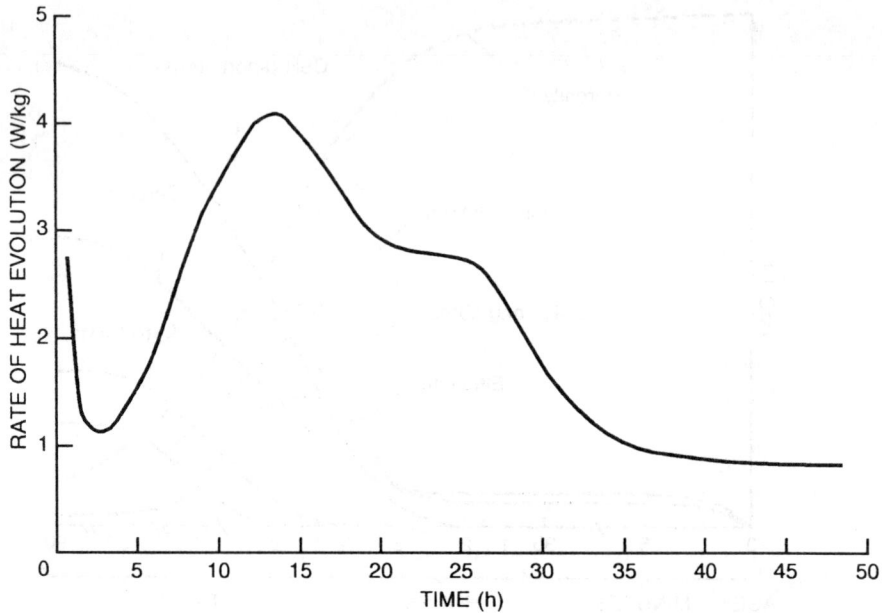

FIG. 4. *Plot of enthalpy change against time for ordinary Portland cement paste [15].*

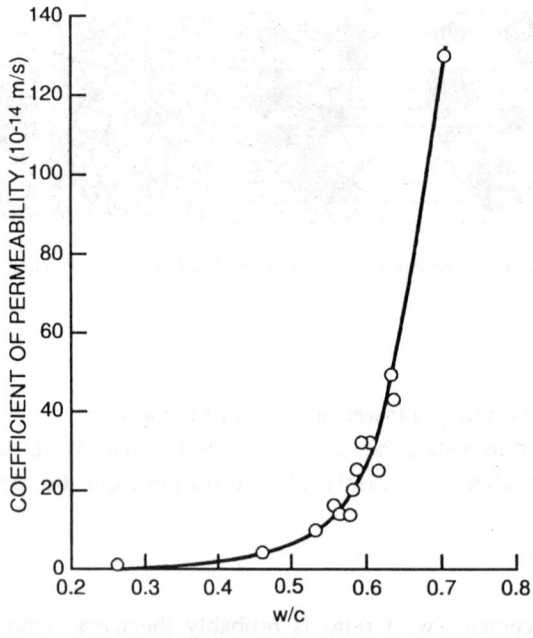

FIG. 5. *Relation between permeability and w/c [16].*

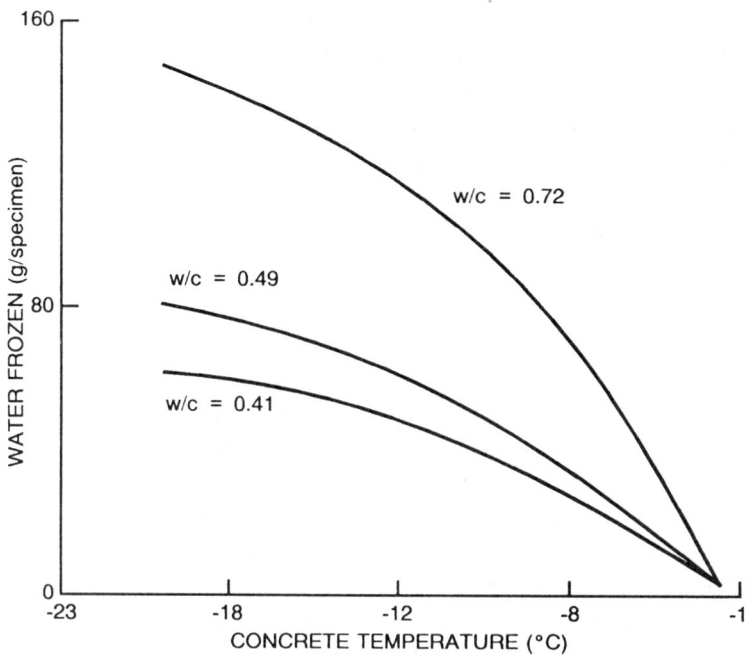

FIG. 6. Relation between amount of water frozen in concrete and temperature at different w/c ratios [17].

preparation of cement paste, substantial excess water is generally required to ensure that the mixture is sufficiently plastic. Excess water not required for hydration may partly evaporate; however, the pores of solidified waste forms remain filled with the aqueous phase.

This aqueous phase is a saturated solution of $Ca(OH)_2$, having a pH above 12. The solubility of $Ca(OH)_2$ in water is about 1.1 g/L. In cements, however, alkalis are also present and they concentrate in the aqueous phase, thereby depressing the solubility of $Ca(OH)_2$. The high pH in unleached cements is thus maintained by the alkali solubility, which in turn depresses the solubility of $Ca(OH)_2$.

Permeability

The higher the w/c ratio, the more pronounced is the permeability. Figure 5 shows the relation between permeability and w/c for mature Portland cement pastes. For w/c above 0.5–0.6, the permeability of the cement paste increases extremely rapidly. Increasing permeability is detrimental to the resistance of concrete exposed to aggressive waters. Calcium hydroxide, formed in the hydration reaction, can be leached quickly from hardened cement. The rate of leaching is governed by the rate of $Ca(OH)_2$ diffusion outward from the surface in contact with water. This action has a strong effect on the movement of radionuclides and other wastes from the waste form, and it is also a slow beginning of corrosion. The leaching of $Ca(OH)_2$ from hydrated cement reduces its CaO content; this leads to the decomposition of the hydrated silicates, aluminates and ferrites, which are not stable unless $Ca(OH)_2$ is present in a certain concentration.

Freeze–thaw resistance

The durability or freeze–thaw resistance of concrete is affected by w/c. Deterioration during freezing and thawing occurs in the pore structure of the cement. As mentioned earlier, voids can be classified as gel pores, capillary pores and air pores. When water gets into hardened cement, both the gel pores and the capillary pores become filled. The gel pores are so small that it is impossible for water to freeze at temperatures higher than $-78°C$. Ice crystals cannot form since no more than a dozen or so molecules of water can occupy the gel pore. In 'frozen' cement mixtures, therefore, water in the gel pores is supercooled but not frozen. Capillary pores, however, are large enough to accommodate ice crystals. One cubic centimetre of water occupies approximately 1.09 cm^3 after freezing, and any void in the concrete which is more than 90% full of water will be subjected to pressure when freezing water turns to ice, unless the excess water can be forced from the void during freezing. Figure 6 shows the amount of water frozen in a hydrated cement specimen as a function of w/c at different temperatures. A proper w/c ratio will lessen the detrimental effects of freeze–thaw on concrete and ensure its durability.

The large void volume resulting from high w/c ratios affects practically all of the mechanical, durability and chemical properties of hardened cement.

3.2.1.4. Chemical properties

Dry concrete is immune to attack by dry chemicals and is highly resistant to some chemicals in solution. However, a considerable number of chemicals in solution may attack concrete. A few such chemicals occur in nature, including acid waters, solutions of sodium or magnesium sulphate, and sea water. When hydrated cement is attacked by a chemical solution, the rate and extent of deterioration depend on the concentration of the substance, the temperature and pressure of the solution and, more importantly, the quality of the cement mixture. Concentrated solutions attack more energetically, higher temperatures increase the rate of attack and solutions under pressure enter the concrete more rapidly.

Sulphate attack

Sulphate ions are mainly responsible for the aggressivity of water. Owing to its wetting power, sulphate containing water can penetrate deeply into the hardened cement. Sodium sulphate (Na_2SO_4), for example, can react with hydrated calcium aluminate in Portland cement to produce a material called ettringite ($3CaO \cdot Al_2O_3 \cdot 3CaSO_4 \cdot 31H_2O$), which has a volume double that of the initial solids. The generation of ettringite causes the hardened cement to expand and crack. Sodium sulphate also reacts with $Ca(OH)_2$ in Portland cement mixtures to produce gypsum ($CaSO_4 \cdot 2H_2O$), which has a volume greater than that of the solids entering the reaction. The increase in volume causes the pores of the hardened cement to expand and burst, particularly those in the surface layer. Ammonium, calcium and magnesium sulphates are also detrimental to Portland cement. Magnesium sulphate not only attacks the hydrated aluminate, but may also attack the hydrated silicates to form gypsum, magnesium hydroxide and silica gel. In general, Type II Portland cement (with 8% tricalcium aluminate) should be used in cases of exposure to sulphate solutions having 150–1000 ppm as SO_4^{2-}. Type V Portland cement (with 5% tricalcium aluminate) should be used when the SO_4^{2-} concentration exceeds 1000 ppm.

Acid attack

Hydrated cement is chemically basic, its pore water having a pH of approximately 13, and is therefore attacked by acids which have pH values of less than 7. With the exception of carbonic acid, free acids are rare in nature. The acidity of groundwater is due to its CO_2 content, to organic acids (humic acid) or to acidic salts. The acidity of water in marshes or peat regions is mainly due to the development of sulphuric acid or sulphurous acid resulting from the oxidation of H_2S or pyrite (FeS_2). Fluctuations in the pH of groundwaters are caused by changes of season and of weather. The hydrogen ion reacts readily with the cement components, particularly with $Ca(OH)_2$, causing decomposition by ion exchange. In concentrations characteristic of natural waters, acids dissolve the carbonate layer at the surface of hardened cement, preventing further carbonation and thereby promoting the leaching of lime. Deterioration occurs when the acids form water soluble salts with the $Ca(OH)_2$. Oxalic acid and phosphorous acid are the exceptions, since their calcium salts are insoluble in water. Sulphur and sulphurous salts may cause expansion as a result of sulphate corrosion. Deterioration occurs earlier with exposure to H_2SO_4 than to HCl or HNO_3, since the aggressive effect of the hydrogen ions is added to that of the SO_4^{2-} ions, which combine with Ca^{2+} ions to form gypsum. The end product of acid action is the silica gel and calcium and aluminium salts of the corresponding acid. The effect of organic acids on hydrated cement is to transform the $Ca(OH)_2$ into soluble salts. The most detrimental organic acids are acetic acid, lactic acid and formic acid.

TABLE IV. EFFECTS OF CHEMICALS ON HARDENED PORTLAND CEMENT [11, 12, 14]

Chemical	Effect on cement
Acids	
Hydrochloric, nitric, sulphuric, sulphurous	Disintegration
Acetic, carbonic	Slow disintegration
Humic	Little or no disintegration
Natural acidic waters	Surface disintegration
Salts and alkalis	
Carbonates of ammonia, potassium, sodium	None
Chlorides of calcium, potassium, sodium, strontium	None unless concrete is wet with chloride solution
Chlorides of ammonia, copper, iron, magnesium, mercury, zinc	Slow disintegration
Hydroxides of ammonia, calcium, potassium, sodium	None
Nitrates of ammonia	Disintegration
Nitrates of calcium, potassium, sodium	None
Sulphates of aluminium, ammonia, calcium, cobalt, copper, iron, magnesium, manganese, nickel, potassium, sodium, zinc	Disintegration

Effect of salts

The effect of salts is essentially identical to but considerably weaker than that of the corresponding acid. The salts of nitric acid, the nitrates, are readily soluble in water, but sodium nitrate ($NaNO_3$), potassium nitrate (KNO_3), calcium nitrate ($Ca(NO_3)_2$), nickel nitrate ($Ni(NO_3)_2$) and lead nitrate ($Pb(NO_3)_2$) do not attack hydrated cements.

Chloride attack

The chlorides are less dangerous than the sulphates, and while some chlorides are chemically aggressive, others are harmless. The harmful chlorides are ammonium chloride (NH_4Cl), mercury chloride (sublimate, $HgCl_2$) and, to a lesser extent, iron chloride ($FeCl_2$) and aluminium chloride ($AlCl_3$). Chlorides react with the $Ca(OH)_2$ by cation exchange to form soluble $CaCl_2$.

Effect of bases

Portland cement is highly resistant to strong solutions of most bases. It is unaffected by continuous exposure to 10% solutions of sodium or potassium hydroxides; calcium, ammonium, barium and strontium hydroxides in the same concentration are also harmless. At concentrations of basic solutions above 20%, corrosion takes place as a result of the dissolution of the silicates and aluminates formed in the hydration of Portland cement.

Table IV summarizes the effects of chemicals on hardened Portland cement.

3.2.1.5. Non-ferrous metals

Non-ferrous metals are frequently used in construction with Portland cement concrete. Experience gained in construction work could be useful in the assessment of interactions of cemented waste forms with the container materials. Copper, zinc, aluminium, lead and alloys containing these metals may be corroded when embedded in or in surface contact with concrete. Corrosion of embedded metals is caused by direct oxidation in strong alkaline solutions that normally occur in fresh cement mixtures such as concrete, mortar and cement paste, or by galvanic currents between two dissimilar metals in the presence of an electrolyte. Copper and copper alloys are practically immune to action from fresh cement mixtures unless the mixtures contain soluble salts.

Zinc is susceptible to attack by fresh cement mixtures. This attack results in the evolution of hydrogen and the formation of calcium zincate, which occupies a greater volume than the original metal and may exert expansive pressures around the embedded element.

Aluminium is attacked when embedded in hardened cement. Initially the reaction results in the formation of aluminium oxide and the evolution of hydrogen. The greater volume occupied by the oxide causes expansive pressures.

Lead is attacked by fresh cement mixtures and is converted to lead oxide or to a mixture of lead oxides. Corrosion continues in the presence of moisture and may totally destroy the lead in a short time.

3.2.2. Blast-furnace slag cements

3.2.2.1. General composition

Blast-furnace slag, a by-product of the iron and steel industry, is a combination of the earthy constituents of the iron ore with limestone flux. The essential components of slag are the same oxides as are present in Portland cement, namely lime, silica and alumina, but their proportions differ.

The extent to which slag is used in cements varies in different countries. Slag is particularly widely used in a number of western European countries, and several different types of cement are produced.

(1) Ground slag, mixed with a suitable proportion of limestone, is used as a raw material for the manufacture of Portland cement.

(2) Granulated blast-furnace slag is ground with Portland cement clinker in various proportions. Such cements include the Portland blast-furnace cements in the United Kingdom and the USA, the low heat slag cement in the UK, the Hochofen cements in Germany, and the Ciment Portland de fer, Ciment métallurgique mixte, Ciment de haut fourneau and Ciment de laitier au clinker in France. In a Belgian development, known as the Trief Process, the granulated slag is ground wet and is added either as a slurry or after drying to the cement and aggregate in the concrete mixer.

(3) Granulated slag is ground with a small proportion of dead burnt gypsum or anhydrite and a smaller amount of cement or lime. Such products are known as supersulphated cements and are made mainly in Belgium but also in France, Germany and the UK.

(4) Ground granulated slag is mixed with hydrated lime. The French and Belgian Ciments de laitier à la chaux and the US slag cement are of this type, which was also formerly made in Germany and the UK, where it was sometimes known as cold process slag cement. Table V shows the compositions of some blast-furnace slags.

TABLE V. COMPOSITION (%) OF BLAST-FURNACE SLAGS [11]

Source	Type	CaO	SiO_2	Al_2O_3	MgO	Fe_2O_3	MnO	S
France	Haematite	40–48	29–36	13–19	2–8	0.5–3.8	0.1–1.0	0.4–1.5
Germany	Thomas	38–46	29–35	10–16	4–12	0.2–1.0	0.1–1.3	0.6–1.9
	Foundry	35–43	30–40	11–16	5–11	0.2–1.3	0.4–2.0	0.9–1.8
	Haematite	40–44	34–35	11–13	6–8	0.3–0.4	0.5–1.1	1.4–1.8
South Africa	Haematite	28–39	28–38	10–22	7–21	0.4–3.0	0.2–0.9	0.7–1.4
Former USSR	Haematite	29–48	34–35	5–23	0.18	0.3–2.4	0.1–2.1	1.1
UK	Basic	37–42	30–36	12–22	3–11	0.3–2.1	0.4–2.2	0.9–1.9
	Foundry	39–44	30–38	15–23	3–8	0.1–1.0	0.2–0.7	1.4–2.5
	Haematite	38–42	32–37	10–22	3–9	0.4–1.3	0.3–1.0	1.1–2.4
USA	Haematite	36–45	33–42	10–16	3–12	0.3–2.0	0.2–1.5	—

The composition of blast-furnace slags depends on the raw materials and industrial processes used, but the slag should always have a high lime content (approximately 40%) when used as a cement. Furthermore, the physical structure of the slag depends on its rate of cooling; for use as a cement, rapid cooling is necessary to quench the material to form a reactive glass and to prevent the crystallization of unreacted chemical compounds. Slag is generally quenched with water, a process known as granulation.

Some slag will hydrate very slowly on contact with water. Its hydration is activated by the addition of other compounds, such as calcium hydroxide, calcium sulphate, sodium carbonate and sodium sulphate.

Blast-furnace slag cements have physical properties similar to those of ordinary Portland cements. The distribution of particle size and the surface area of blast-furnace slags depend on the method of manufacture, but in general their fineness is similar to that of Portland cements [17].

3.2.2.2. Types

Portland blast-furnace cement

Lime for activation is most conveniently supplied by the hydration of Portland cement. Table VI shows that wide variations in slag content are allowable in Portland blast-furnace cements. These cements are widely used in Europe but are not popular in the USA. The amounts of slag and Portland cement used vary from country to country; in the USA, ASTM C595-79 for Type IS cements allows 25–65% slag to be blended with Portland cement. British Standard (BS) 146:1973 allows any content up to 65%, while BS 4246:1974 specifies a slag content of 50–90%. This wide variation in slag content means that properties of the cement vary widely. Slags react slowly to form CSH (calcium silicate hydrate), the same hydration product as is

TABLE VI. TYPES OF SLAG CEMENT [12]

Description	Activator	Percentage slag	Comments
Lime slag cement	CH	70	Obsolete
Portland blast-furnace slag cement	CH	20–85	CH from Portland cement hydration
Supersulphated cement	CS	80–85	Small amount of Portland cement added

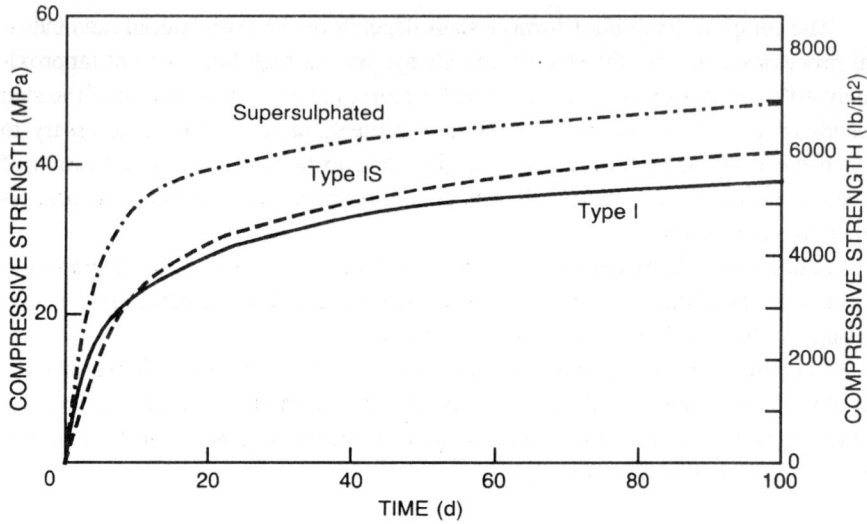

FIG. 7. *Compressive strength development of concretes made with different slag cements* [12].

formed by the hydration of C_3S and C_2S. Thus, early strength development is slower than that of Portland cement (Fig. 7), depending on the amount of slag present, and the heats of hydration are also lower. Slag cements with high slag contents generally have better sulphate resistance than a Type I Portland cement, probably owing to a lower calcium hydroxide content. Thus Type IS cements have properties comparable to those of a Type IV Portland cement [14].

Supersulphated slag cement

When slag is activated with calcium sulphate in the form of anhydrite ($CaSO_4$), together with a small amount of lime or Portland cement, the product is known as supersulphated cement. This is not strictly speaking a modified Portland cement, but it forms similar hydration products. The cement is not available in the USA but is produced in Europe (BS 4248:1974, for example), although it is less widely used than slag cements. Supersulphated cements have lower heats of hydration than most Portland blast-furnace slag cements and generally show better resistance to sulphate. This resistance results from a low CH content and also because most of the alumina stays combined as ettringite. It is believed that ettringite is responsible for much of the early strength of a supersulphated cement (although it is not a rapid hardening cement) [14].

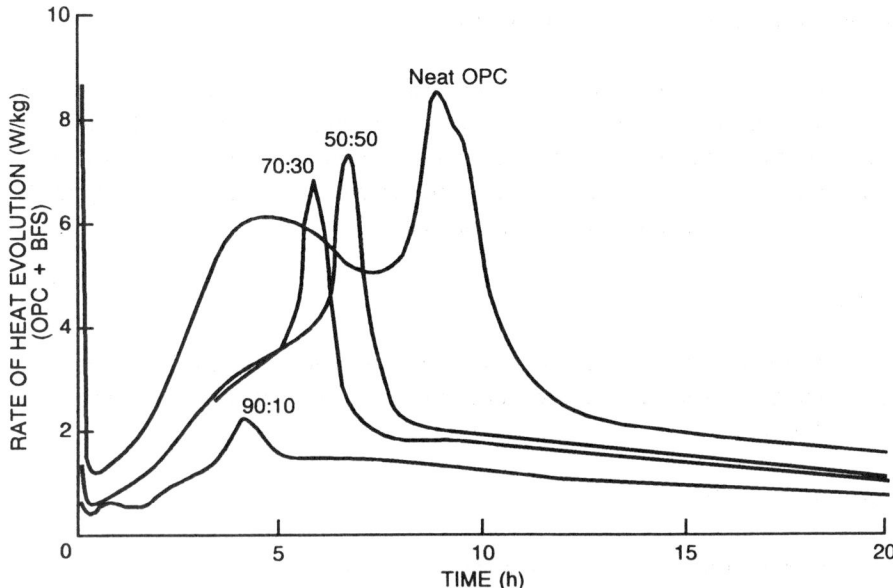

FIG. 8. *Time dependence of heat evolution rate for various ratios of ordinary Portland cement to blast-furnace slag cement [18].*

3.2.2.3. Hydration

The hydration of blast-furnace slag cements proceeds in a similar manner to that of Portland cements. The major differences are in the rate of heat evolution, the total heat of hydration and the relative amounts of hydration products.

As more of the Portland cement in blast-furnace slag–Portland cement blends is replaced by blast-furnace slag, the rate of heat evolution decreases (Fig. 8 [18]). This is an important consideration in the full scale application to the cementing of radioactive wastes, where 75–90% blast-furnace slag is used to reduce the curing exotherm to less than 100°C, and preferably to less than 80°C [19].

By incorporating at least 75% blast-furnace slag in blast-furnace slag–Portland cement blends, it is possible to virtually eliminate calcium hydroxide as a hydration product; this is an important consideration in immobilizing certain types of waste such as sulphates and ion exchange resins. Differences in the aluminate hydrates may also contribute to the improved chemical resistance of blast-furnace slag cements.

3.2.2.4. General properties

In addition to reduced heat of hydration, other benefits are derived from the use of blast-furnace slag.

Permeability and distribution of pore size

Despite an increase in total porosity, blast-furnace slag cements have a lower permeability than equivalent Portland cements, owing to the increased volume of smaller pores. Lower permeability contributes to improved resistance to frost, lower diffusion rates of ions through the hardened cement and improved stability in the presence of salts, such as chloride and sulphate [20].

Setting rate

The reduced setting rate of blast-furnace slag cements is beneficial in producing cement grouts with an extended working time, for example in grouting processes for solid wastes.

Chemical insolubility

The pore water of blast-furnace slag cements contains hydroxide anions as in Portland cements, but blast-furnace slag cements also contain sulphur species which give them a high pH and low oxidation potential. These properties are beneficial in reducing the solubility of most radionuclides, particularly the actinides, and in reducing the rate of corrosion of the steel containers.

Generally, other physical and mechanical properties of blast-furnace slag cements, such as density and compressive strength, are similar to those of equivalent Portland cements.

3.2.3. High alumina cement

High alumina cement is obtained by fusing or sintering a suitable mixture of aluminous and calcareous materials and grinding the product to a fine powder.

The raw materials used to manufacture high alumina cement are limestone and bauxite [11].

Table VII gives typical values of oxide composition of high alumina cement. A minimum alumina content of 32% is prescribed by BS 915:1972, which also requires the alumina/lime ratio to be between 0.85 and 1.3 [2].

High alumina cement, which is composed primarily of monocalcium aluminate ($CaO \cdot Al_2O_3$), is a rapid curing, high early strength cement that combines with significantly more water during hydration (up to 50 wt%) than Portland cement [7]. Hydration occurs by the following generic reaction:

$$3CaO \cdot Al_2O_3 + (18 + x) \cdot H_2O = CaO \cdot Al_2O_3 \cdot 10H_2O + 2CaO \cdot Al_2O_3 \cdot 8H_2O$$
$$+ Al_2O_3 \cdot xH_2O(aq)$$

TABLE VII. TYPICAL OXIDE COMPOSITION OF HIGH ALUMINA CEMENT [14]

Oxide	Content (%)
SiO_2	3–8
Al_2O_3	37–41
CaO	36–40
Fe_2O_3	9–10
FeO	5–6
TiO_2	1.5–2
MgO	1
Insoluble residue	1

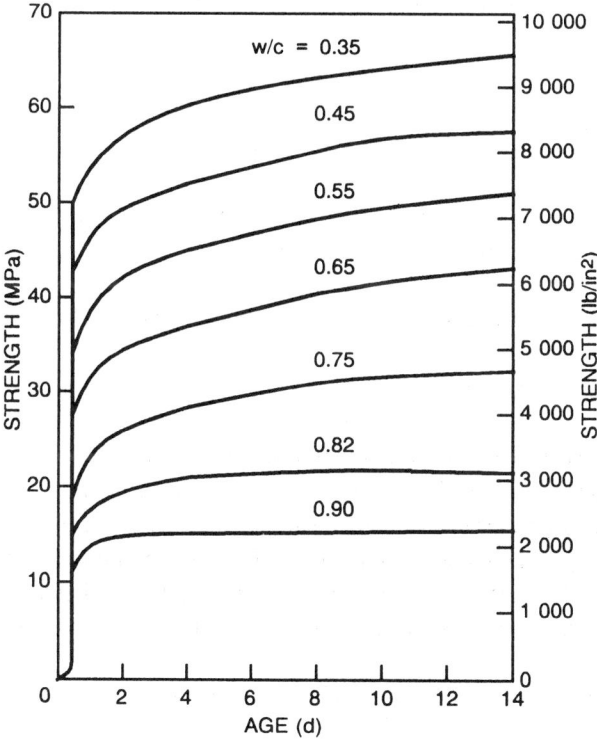

FIG. 9. Strength development of high alumina cement with different w/c ratios (cured at 18°C and 95% r.h.) [22].

TABLE VIII. COMPOSITION (%) OF HIGH ALUMINA CEMENT [11]

Source	Type of manufacture	SiO_2	Al_2O_3	CaO	Fe_2O_3	FeO	TiO_2	MgO	S^{2-}	SO_3^{2-}
Czechoslovakia	Brick kiln type furnace sintering	6–8	40–45	37–42	12–14	Trace	<2	1	Trace	0.5
France	Reverberatory furnace fusion	3.5–4.5	38–40	36–39	9–11	4–6	<2	1	Trace	0.1
Germany	Blast-furnace reductive fusion	5–8	48–51	39–42	0.1	<1	1.5	1	1	0.5
Spain	Reverberatory furnace fusion	4–5	36–38	39–42	10–12	4–5	<2	1	Trace	0.1
UK	Reverberatory furnace fusion	4–5	38–40	36–39	8–10	5–7	<2	1	Trace	0.1
USA	Rotary kiln fusion	8–9	40–41	36–37	5–6	5–6	<2	1	0.2	0.2
Yugoslavia	Reverberatory furnace fusion	6–8	38–40	36–39	8–10	4–7	<2	1	Trace	0.1

The decahydrate ($CaO \cdot Al_2O_3 \cdot 10H_2O$) is unstable at normal room temperatures and is ultimately transformed into the hexahydrate and alumina gel [21]:

$$3CaO \cdot Al_2O_3 \cdot 10H_2O = 3CaO \cdot Al_2O_3 \cdot 6H_2O + Al_2O_3 \cdot H_2O(aq)$$

3.2.3.1. Physical properties

The influence of w/c on the strength of high alumina cement cured at 18°C for 14 days is shown in Fig. 9. The high strength of the cement is reached when the hydration of calcium aluminate (CA) results in the formation of hydrated calcium aluminate (CAH_{10}), with a small quantity of dicalcium aluminate (CA_2AH_8) and of alumina gel ($Al_2O_3(aq)$). However, the hydrate CAH_{10} is chemically unstable at normal temperatures and transforms into hydrated tricalcium aluminate (C_3AH_6) and alumina gel.

Table VIII shows the composition of some high alumina cements originating in various countries.

The loss of strength of cold cured cement is smaller in mixes with low w/c than in mixes with high w/c (Fig. 10) [14, 22].

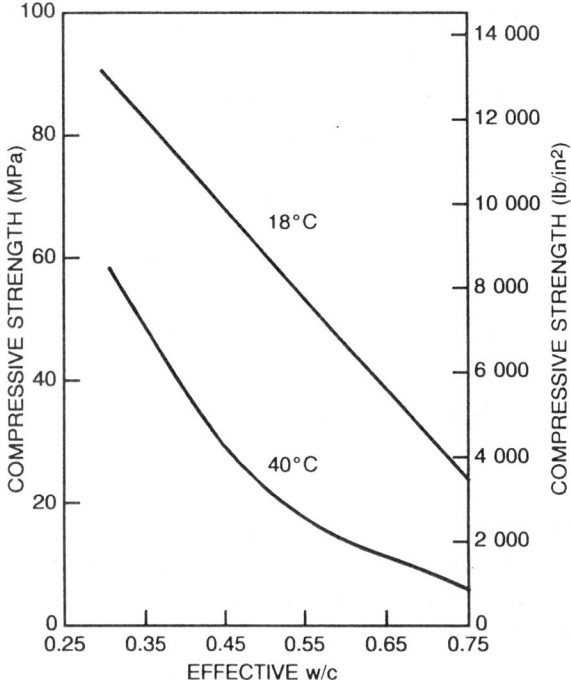

FIG. 10. *Influence of w/c on the strength of high alumina cement cured in water at 18 and 40°C for 100 d [22].*

3.2.3.2. Chemical properties

Sulphate attack

High alumina cement was first developed to resist sulphate attack. This resistance to sulphates may be due to the absence of $Ca(OH)_2$ in hydrated high alumina cement and also to the protective influence of the relatively inert alumina gel formed during hydration [23, 24].

The relative resistance of Portland and high alumina cements to sulphates is shown in Table IX, which gives the rates of expansion of mortars immersed in various solutions.

Acid attack

High alumina cement is resistant to waters which are acidic owing to the presence of free carbonic acid, as in some waters containing very little dissolved solids. The cause of the increased resistance of high alumina cements to such acidic

27

TABLE IX. LINEAR EXPANSION (%) OF 1:3 CEMENT MORTARS IN SULPHATE SOLUTIONS [11]

Solution	Portland cement				High alumina cement
	4 weeks	12 weeks	24 weeks	1 year	
5% Na_2SO_4	0.018	0.070	0.144	0.32	No expansion in 1 year
5% $MgSO_4$	0.018	0.054	0.25	0.91	No expansion in 1 year
5% $(NH_4)_2SO_4$	0.100	3.800	—	—	No expansion in 1 year

waters is uncertain, but the presence of alumina gel enveloping the more susceptible lime compounds is important. The cement also remains intact in dilute solutions of organic acids (pH greater than 3.5–4.0) found in waste effluents. However, it is less resistant to strong solutions of hydrochloric, hydrofluoric and nitric acids [24].

Effect of alkaline solutions

Even dilute solutions of caustic alkalis attack high alumina cement with great vigour by dissolving the alumina gel [24].

Although high alumina cement stands up extremely well in sea water, sea water should not be used for mixing since it adversely affects the setting and hardening of the cement, possibly because of the formation of chloroaluminates. Likewise, calcium chloride must not be added to high alumina cement [14].

3.2.4. Masonry cements (high lime cements)

Masonry cement is a mixture of Portland cement and slaked lime, $Ca(OH)_2$. In radioactive waste solidification, Portland cement and slaked lime are typically combined in equal proportions. When water is present, the extremely high alkalinity induced by the slaked lime produces a rapid setting cement. Masonry cement is particularly useful for solidifying wastes, such as boric acid and borate salts, that tend to inhibit or retard hydration. Masonry cement also has advantages over Portland cement in solidifying organic liquid wastes [25], presumably owing to the increased rate of cement hydration reactions caused by the alkalinity of slaked lime. The bulk density of masonry cement is about 35% less than that of Portland cement. In some cases, masonry cement can incorporate more waste than Portland cement [26]. While substantial, the compressive strength of masonry cement paste is significantly less than that obtained with Portland cement under similar conditions [27]. The high concentration of slaked lime in masonry cement compared with ordinary Portland cement allows significantly higher amounts of boric acid waste to be incorporated into a waste form. The major reactions that occur between slaked lime and boric acid are:

$$2H_3BO_3 + Ca(OH)_2 \longrightarrow Ca(BO_2)_2 + 4H_2O$$
$$2H_3BO_3 + Ca(OH)_2 + 2H_2O \longrightarrow Ca(BO_2)_2 \cdot 6H_2O$$
$$4H_3BO_3 + Ca(OH)_2 \longrightarrow CaB_4O_7 + 7H_2O$$

The end products of these chemical reactions are calcium metaborate, calcium metaborate hexahydrate and calcium tetraborate, respectively. These compounds are insoluble or only slightly soluble and therefore do not ionize well.

Figure 11 shows a ternary compositional phase diagram for cement waste forms containing various concentrations of boric acid waste in masonry cement. The figure shows that approximately 15 wt% boric acid can be incorporated into masonry

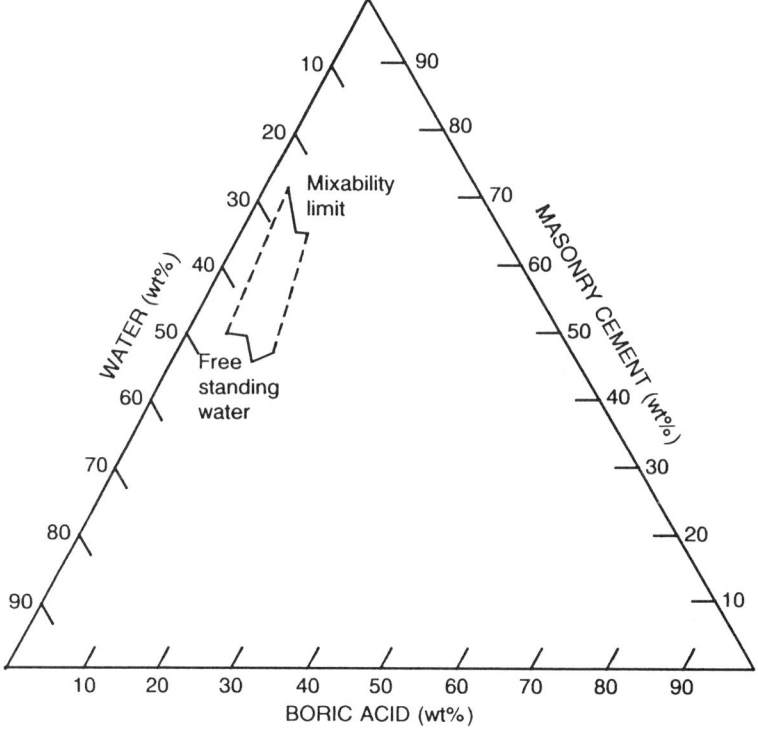

FIG. 11. *Ternary compositional phase diagram for the solidification of boric acid wet evaporator bottoms with masonry cement [27]. Broken line indicates range of boric acid liquid waste concentration tested.*

29

cement owing to its increased lime content. In contrast, as little as 5 wt% boric acid will inhibit the curing of ordinary Portland cement. Formulations containing 15 wt% boric acid have a molar ratio of acid to base approximately equal to 1. Above 15 wt% boric acid the inhibiting properties of boric acid dominate.

3.2.5. Special cements

3.2.5.1. Polymer modified gypsum cements

A polymer modified gypsum cement, Envirostone, that is blended as a powder with liquid wastes has been used to solidify wet solid and organic wastes. This proprietary product can be mixed much like cement, but it requires acid conditions to solidify. Gypsum is a naturally occurring hydrated form of calcium sulphate ($CaSO_4 \cdot 2H_2O$). Gypsum cements are dehydrated gypsum products; the most common is plaster of Paris ($CaSO_4 \cdot \frac{1}{2}H_2O$). Gypsum cements combine with water to form a solid mass of gypsum:

$$CaSO_4 \cdot \tfrac{1}{2}H_2O + \tfrac{3}{2}H_2O \longrightarrow CaSO_4 \cdot 2H_2O$$

During the manufacture of Envirostone, calcium sulphate hemihydrate is blended with a water dispersible melamine–formaldehyde resin which is hydrophobic when cured (GB Patent 2 097 990A). Linking of the resin is started by the water content of the wastes in conjunction with an amino based cross-linking agent which is also incorporated in manufacture. Envirostone has been tested with a wide range of wastes, e.g. concentrated boric acid solutions, bead and powdered ion exchange resins and decontamination solutions [28]. It can also be used to immobilize oil with the aid of suitable non-ionic surfactants (GB Patent 2 137 403A). The strength of waste forms prepared with Envirostone is low in comparison with those with Portland cement, but high waste loadings can readily be achieved since the side reactions which often interfere with the hydration of ordinary Portland cement are avoided. Processing times are short and can be adjusted, if necessary, by the use of retarders.

3.2.5.2. Expansive cements

One of the major disadvantages of ordinary Portland cement concrete is its high shrinkage on drying and its susceptibility to tensile cracking when contraction is wholly or partially restrained (Fig. 12(a)). Shrinkage cracking destroys the integrity of the concrete and is unsightly; thus special allowance for this must be made in design and construction. On the other hand, volume expansions during moist curing in ordinary Portland cements are very small. If there were greater expansion during hardening, this could offset the contraction that occurs on drying (Fig. 12(b)). The development of expansive cements dates back about fifty years. Commercial

30

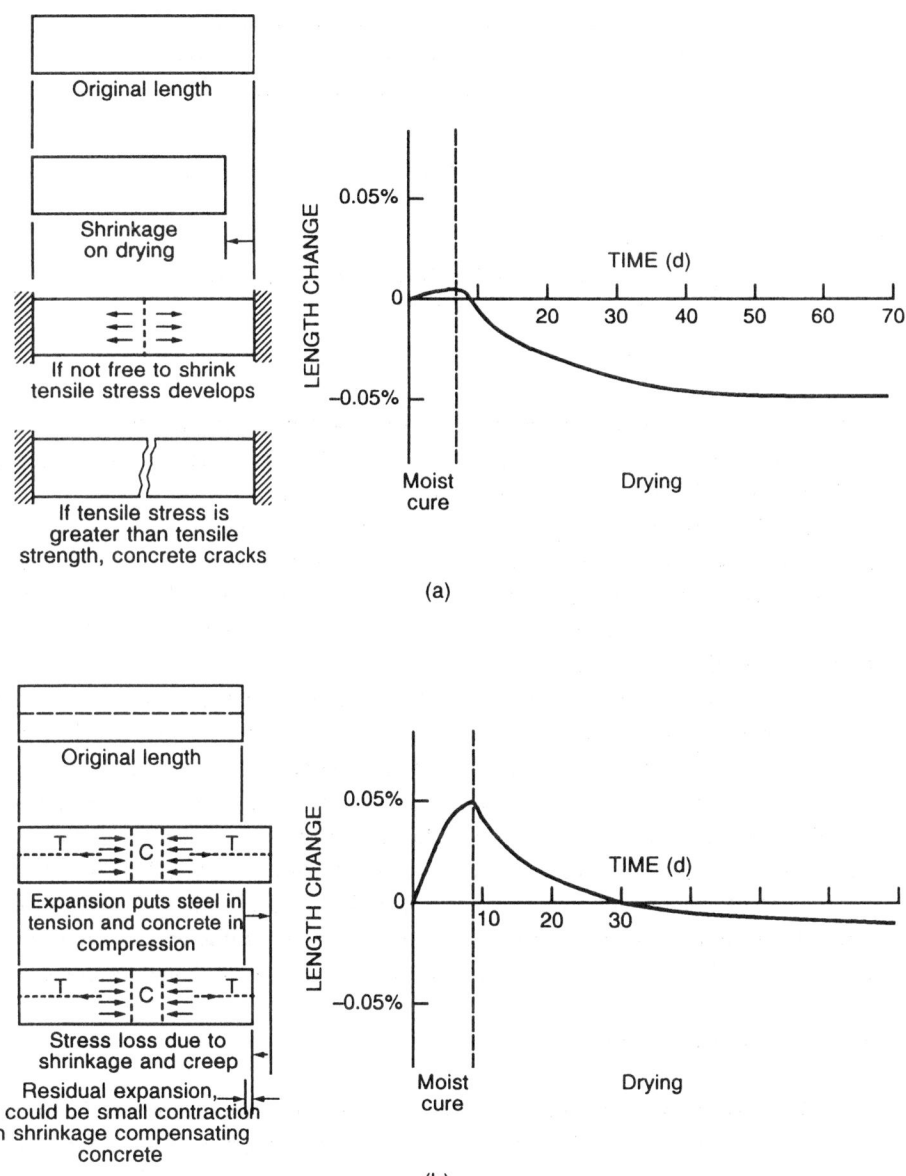

Original length

Shrinkage
on drying

If not free to shrink
tensile stress develops

If tensile stress is
greater than tensile
strength, concrete cracks

LENGTH CHANGE

0.05%

0

-0.05%

TIME (d)

20 30 40 50 60 70

Moist
cure

Drying

(a)

Original length

T ⇒ C ⇐ T

Expansion puts steel in
tension and concrete in
compression

T ⇒ C ⇐ T

Stress loss due to
shrinkage and creep

Residual expansion,
or could be small contraction
in shrinkage compensating
concrete

LENGTH CHANGE

0.05%

0

-0.05%

TIME (d)

10 20 30

Moist
cure

Drying

(b)

FIG. 12. *Drying shrinkage of concretes made with (a) Type I Portland cement and (b) expan-*
sive cement [29].

TABLE X. COMPOSITION OF COMPONENTS
IN EXPANSIVE CEMENTS [12]

Type K	Type M	Type S
C_4A_3S	CA	C_3A
CS	$(C_{12}A_7)^a$	CSH_2
$(C)^a$	CSH_2	

[a] Generally contained in the expansive components but not necessary for expansion.

production began in the USA in the late 1960s, although total production remains quite small, about 500 000 tonnes annually. A tentative standard specification, ASTM C845-76T, covers expansive cements [30, 31].

All three variants of expansive cements currently in use are based on the formation of considerable quantities of ettringite during the first few days of hydration. The materials from which ettringite is formed differ substantially in each cement (Table X), but all require a source of calcium aluminate and sulphate ions: (CA) + S + H → ettringite.

Type S is basically a Type I Portland cement with a high C_3A content. Type K contains the anhydrous calcium sulphoaluminate C_4A_3S, which can be formed in the rotary kiln as an integral part of the cement. Alternatively, C_4A_3S can be formed separately and blended with a Type I cement together with the necessary anhydrite (CS). Lime is not necessary for expansion but it is often present and changes the early rate of expansion. The aluminate source of Type M is the primary constituent of calcium aluminate cement and is blended with a Type I cement. Several expansive admixtures that can confer expansive characteristics on a Type I cement when the concrete is being proportioned [12] are available commercially.

3.2.5.3. Rapid setting cements

The formation of large quantities of ettringite is also responsible for the properties of two special cements that develop considerable strength within 1–2 hours after casting. These are regulated set cement (or jet cement) and very high early strength cement (VHE cement). Regulated set cement, developed in the 1960s, is not currently produced in North America but is being manufactured in Japan and Germany. However, VHE cement is commercially available in North America. In contrast to expansive cements, in VHE cements the formation of ettringite gives rise to rapid development of strength, and expansions are not as great. In these cements, ettringite

TABLE XI. TYPICAL COMPOSITIONS (%) OF VHE AND REGULATED SET
CEMENTS [12]

Cement compounds	VHE cement	Regulated set cement	Type I ordinary Portland cement[a]
C_3S	—	60	50
C_2S	50	5	25
C_4AF	20	5	8
C_3A	—	—	12
C_4A_3S	20	—	—
$C_{11}A_7 \cdot CaF_2$	—	20	—
Total SO_3	10	10	3

[a] For comparison.

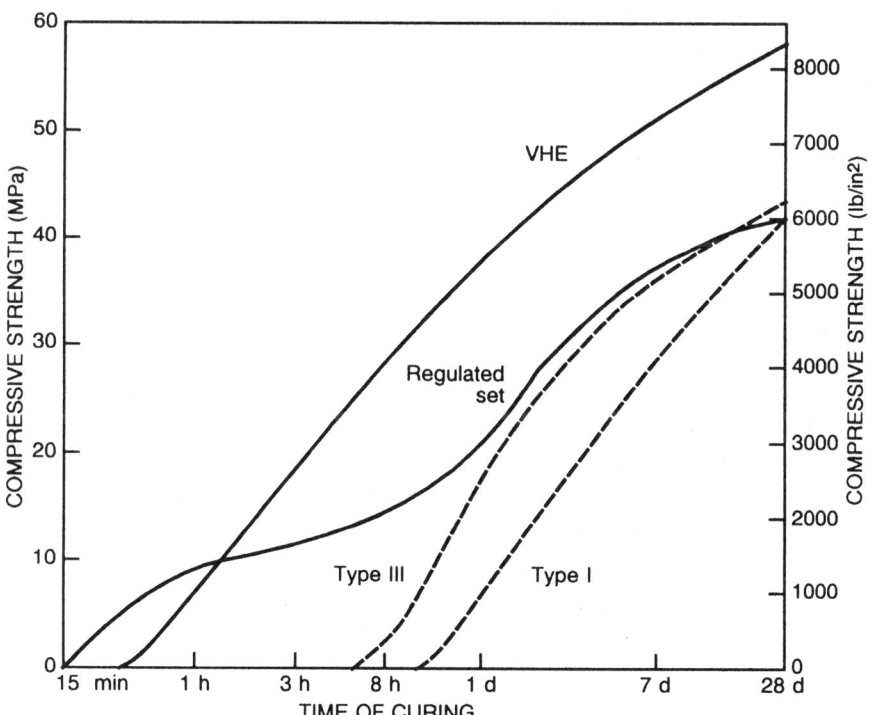

FIG. 13. *Strength development of concretes made with rapid hardening cements [32].*

formation is much more rapid, occurring largely before the paste has gained strength and well before the hydration of the calcium silicates.

Regulated set cements

Regulated set cement is a modified Portland cement in which C_3A is replaced by a new compound, calcium fluoroaluminate. The cement can be produced directly in a rotary kiln or the fluoroaluminate can be blended with a Type I clinker. Calcium fluoroaluminate is more reactive with water than is C_3A and hydrates so rapidly that flash setting always occurs unless sulphate ions are present. The reaction with sulphate ions to form ettringite is also very vigorous, so much so that gypsum will not retard setting because it cannot dissolve fast enough to maintain an adequate supply of sulphate ions. The use of a more soluble salt, such as calcium sulphate hemihydrate (plaster) or sodium sulphate, is needed to avoid flash setting. The time of setting (handling time) can be controlled from about 2 to 40 min using a soluble sulphate, and can be further extended with an organic retarder such as citric acid. The subsequent supply of sulphate ions required to form additional ettringite to develop strength is maintained by slightly soluble anhydrite $(C\overline{S})$.

Strength develops very rapidly after setting occurs, and strengths of 7 MPa can be attained within 1 hour. The initial rise in strength is due to ettringite forming from calcium fluoroaluminate. Once this reaction slows down, development of strength also slows down, until C_3S begins to hydrate. The level of the early strength plateau depends on the amount of fluoroaluminate used, which can be controlled either by the cement content of the concrete or by the fluoroaluminate content of the cement.

Very high early strength cements

In the production of VHE cement, calcium sulphate is added to the raw mix so that $C_4A_3\overline{S}$ is formed in the rotary kiln. This compound is present in Type K expansive cements, but the quantities are greater in VHE cement. Calcium sulphate is also formed or is added during grinding. A typical composition is given in Table XI, which shows that C_2S rather than C_3S is the calcium silicate compound.

As in the case of regulated set cement, the setting time of VHE cement can be controlled in the range 2 to 45 min. Strength development is initially similar to that of regulated set cement (Fig. 13), and strengths in excess of 7 MPa can be attained with 1 h of set. Later strengths are higher than for regulated set cement or Types III and I. As with regulated set cement, the formation of ettringite is accompanied by a high rate of heat evolution, but if this early heat can be dissipated, the subsequent hydration of C_2S will not cause an additional rise in temperature. The presence of C_2S improves the durability of the cement paste because of the lower content of calcium hydroxide in the mix, but the long term sulphate resistance of the cement may not be high. If the concrete is air entrained (air bubbles are incorporated into the cement), frost resistance should be satisfactory. Creep and drying shrinkage are reported to be lower than for Type III cement concrete.

In the UK, an extra rapid hardening cement is sold which is a Type III cement interground with calcium chloride to accelerate hydration. In Japan, superhigh early strength cement has been developed with C_3S contents exceeding 70% and traces of chromium, manganese and fluoride, which are said to increase the rate of hardening [33].

4. CEMENT ADMIXTURES

4.1. INTRODUCTION

A cement admixture is a material other than water, aggregate and hydraulic cement that is used as an ingredient before or during mixing. Admixtures encompass a wide range of specifically formulated chemicals, used in small, predetermined amounts where good quality control can be exercised. The term additive belongs to an older terminology in which it was used synonymously with the term admixture. An additive is a material that is intergrained or blended in limited amounts into a hydraulic cement during manufacture either as a 'processing addition' to aid in manufacturing and handling the cement or as a 'functional addition' to modify the properties of the finished product.

Admixtures are broadly categorized as follows:

(a) *Air entraining agents* are added to improve the frost resistance of concrete.
(b) *Chemical admixtures* are water soluble compounds added to control setting and early hardening of fresh concrete, or to reduce its water requirements.
(c) *Mineral admixtures* are finely divided solids added to cement mixes to improve their workability or durability, or to provide additional cementing properties. Slags and pozzolans are important categories of mineral admixtures.
(d) *Miscellaneous admixtures* include all those materials that do not come under any of the above categories; many of them have special applications.

Since almost every property of concrete can be partly modified, there are a large number of products marketed as admixtures for cement [34]. Table XII shows several properties of concrete that can be improved with various admixtures.

The need for careful selection of cement, waste type and proportioning, and for effective mixing and curing is not lessened by the use of an admixture. The proper amounts are recommended by the manufacturer but tests should be run to check that the desired effects are being obtained.

Table XIII gives the active chemicals in the main admixtures.

TABLE XII. BENEFICIAL EFFECTS OF ADMIXTURES ON PROCESSING OPERATIONS AND PROPERTIES OF CONCRETE [35]

Property	Type of admixture	Category of admixture
Workability	Water reducers	Chemical
	Air entraining agents	Air entraining
	Inert mineral powder	Mineral
	Pozzolans	Mineral
	Polymer latexes	Miscellaneous
Set control	Set accelerators	Chemical
	Set retarders	Chemical
Strength	Water reducers	Chemical
	Pozzolans	Mineral
	Polymer latexes	Miscellaneous
	Set retarders	Chemical
Durability	Air entraining agents	Air entraining
	Pozzolans	Mineral
	Water reducers	Chemical
	Corrosion inhibitors	Miscellaneous
	Water repellent admixtures	Miscellaneous
Special concrete properties (superplastic, waterproof)	Polymer latexes	Miscellaneous
	Slags	Mineral
	Expansive admixtures	Miscellaneous

4.2. AIR ENTRAINING AGENTS

Extensive laboratory testing and long term field experience have demonstrated conclusively that concrete must be properly air entrained if it is to resist freezing and thawing when critically saturated [36, 37]. Figure 14 shows the effect of entrained air on the resistance of concrete to cycles of freezing and thawing [38].

Generally, a loss of strength of 10–20% can be anticipated for most air entrained concretes, particularly those with moderate to high cement contents. The reduction is generally proportional to the amount of air entrained, but there is an

TABLE XIII. CHEMICAL CONTENT OF ADMIXTURES [35]

Type of admixture		Ingredients
Accelerators		Calcium formate, calcium chloride, barium chloride, aluminates, soluble nitrites, triethanolamine, sodium and potassium hydroxides
Retarders		Lignosulphonic acid and salts, hydroxycarboxylic acid and salts, unrefined calcium lignosulphonate, wood resin derivative, phosphates, carbonates, stearate, alkaline aluminates
Water reducers	Normal	Lignosulphonic acid and salts, hydroxycarboxylic acid and salts, hydroxycarboxylic polymer, refined calcium lignosulphonate, sodium lignosulphonate
	Accelerating	Calcium formate and surfactant, calcium chloride and surfactant, calcium chloride and lignosulphonate, calcium chloride and hydroxycarboxylic acid, polyhydroxy compounds, triethanolamine
	Retarding	Lignosulphonic acid and salts, hydroxycarboxylic acid and salts, polyhydroxylic acid and phosphate, hydroxycarboxylic polymer, polyhydroxy compound, phosphate, surfactant
Air entraining agents		Neutralized wood resin, resin and polymer, stearic acid, lignosulphonate, surfactant, animal and vegetable fat, alkyl sulphonate, hydroxycarboxylic acid, sulphonated hydrocarbons, naphthalene sulphonates, sulphonated lignin
Superplasticizers		Sulphonated melamine–formaldehyde condensate, sulphonated naphthalene–formaldehyde condensate, modified lignosulphonate, melamine–formaldehyde, naphthalene sulphonate
Waterproofing agents		Lignosulphonate, stearate, sulphonated carbohydrate, silicones and metal soap, melamine–formaldehyde resin, colloidal silica, organic ester, inorganic silicates, calcium lignosulphonate

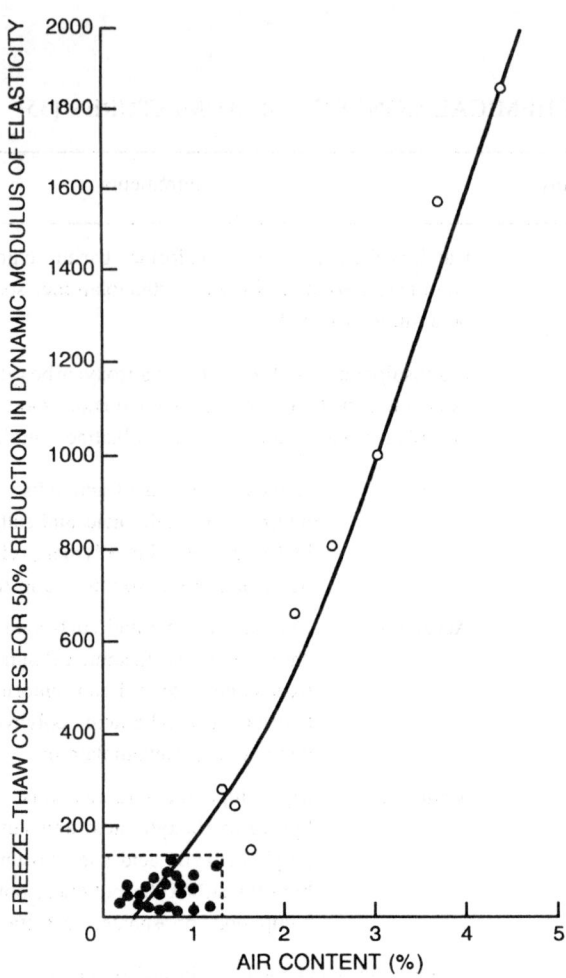

FIG. 14. *Effect of entrained air on the resistance of concrete to freeze–thaw cycles;* ● *not air entrained,* ○ *air entrained [38].*

optimum amount (5–6%) of entrained air for good durability of concrete [39]. The effect of entrained air on durability is shown in Fig. 15.

Air entrainment alters the properties of unhardened cement mixes. These changes must be provided for in proportioning the mix [40, 41]. Air entrained cement is considerably more workable and cohesive than non-entrained cement with an equal w/c ratio. Segregation and bleeding are reduced. The mechanism of entrainment in a cement mix has been discussed in Refs [42, 43]; it is beyond the scope of this report.

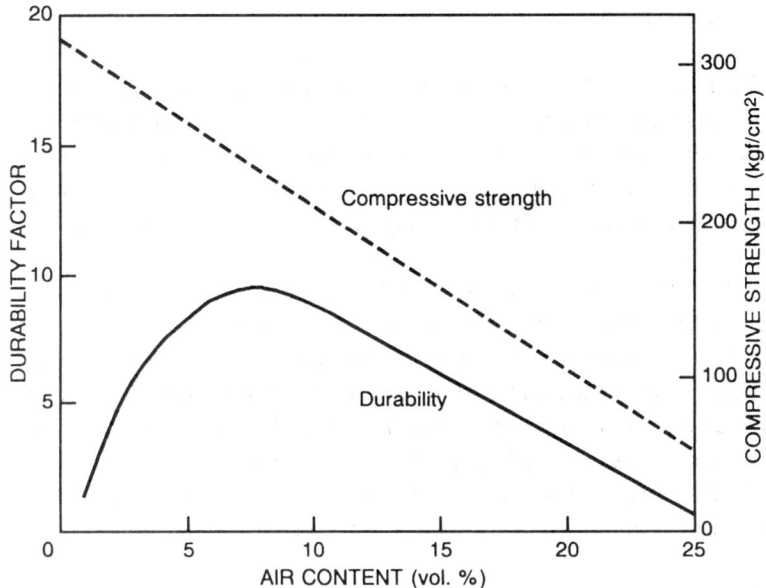

FIG. 15. Effect of entrained air on durability of concrete [39].

The commercial air entraining agents make up a comparatively small group of surfactants. They have been categorized [44] as follows:

— Salts of wood resins
— Synthetic detergents
— Salts of sulphonated lignin
— Salts of petroleum acids
— Salts of proteinaceous materials
— Fatty and resinous acids and their salts
— Organic salts of sulphonated hydrocarbons.

These admixtures create disconnected bubbles which act like ball bearings, evenly distributed and of consistent shape and size.

Adding the admixture at the mixer is preferable because the air content can be controlled within close limits and can be changed readily as required by the work.

4.3. CHEMICAL ADMIXTURES

This class of admixtures encompasses the spectrum of soluble chemicals that are added to concrete to modify setting times and reduce the water requirements of cement mixes.

4.3.1. Accelerators

The early strength of concrete can be increased and the time of setting short-ened by using an accelerating admixture. The decision to use an accelerator is often based on economics. Frequently the same results may be obtained by other means, such as the use of high early strength cement, the use of additional cement, the use of a longer or a different method of curing, heating the water and waste, or a combi-nation of these.

Admixtures which accelerate the hardening of cement mixtures can con-veniently be divided into three groups: soluble inorganic salts, soluble organic com-pounds and miscellaneous solid materials. Numerous investigations have shown that inorganic accelerators act primarily by accelerating the hydration of tricalcium sili-cate [45]. In the case of triethanolamine, Refs [46, 47] indicate that the hydration of tricalcium aluminate is accelerated, but the hydration of tricalcium silicate is retarded. Other organic accelerators may act in a similar way. Quick setting admix-tures probably promote the flash setting of tricalcium aluminate.

4.3.1.1. Soluble inorganic salts

A wide range of soluble inorganic salts, such as chlorides, bromides, fluorides, carbonates, nitrates, thiosulphates, silicates, aluminates and alkali hydroxides, will accelerate the setting and early hardening of Portland cement [48–52]. Calcium chlo-ride is the most widely used accelerator since it is the most effective salt on a weight basis and is more economical. The effects of calcium chloride on the properties of hardened paste, mortar and concrete have been widely studied. Calcium chloride reduces the resistance of concrete to sulphate attack and increases the interaction between high alkali cement and reactive waste components.

4.3.1.2. Soluble organic salts

The most common accelerators in this class are triethanolamine and calcium formate, which offset the retarding effects of water reducing admixtures and provide non-corrosive accelerators. Accelerating properties have been reported for calcium acetate [53], calcium propionate [54] and calcium butyrate, but salts of high carboxylic acid homologues are retarders [50].

Several organic compounds accelerate the setting of Portland cement when low w/c ratios are used [55]. The most notable examples are sugars, which are generally recognized as strong retarders but which promote quick set properties at additions to the cement of greater than 0.25 wt% and at w/c ratios of 0.22–0.24. However, the quick set effect of sucrose may not accelerate gain in strength [46]. Other organic accelerators include urea [50], oxalic acid, lactic acid, various ring compounds and condensation compounds of amines and formaldehyde [51, 56].

Strengths of concrete can be variously decreased or increased by organic accelerators. Both triethanolamine and calcium formate affect the drying shrinkage of concrete in much the same way as does calcium chloride [57].

4.3.1.3. Miscellaneous solid admixtures

It has been reported that the time of setting of Portland cement may be shortened, to varying degrees, by the use of calcium aluminate cement. The compressive strengths at one day or more of neat cement, mortar or concrete prepared with mixtures of Portland and calcium aluminate cements will generally be substantially lower than those obtained with either of the two cements alone. Drying shrinkage and swelling in water are higher for such mixtures and their durability may be reduced [58]. The 'seeding' of Portland cement concrete with 2 wt% finely ground, fully hydrated cement has been reported to be equivalent to the use of 2% calcium chloride, with the additional advantage of increasing the 90 day compressive strength by 20-25% with no increase in drying shrinkage [50, 59-61]. The effects of seeding in addition to the use of calcium chloride are said to be supplementary.

Various silicate minerals act as accelerators [62-64]. Finely divided silica gels and soluble quaternary ammonium silicates were found [65] to accelerate strength development, presumably through the acceleration of tricalcium silicate hydration [52, 66]. Very finely divided magnesium carbonate has been proposed for accelerating the setting times of hydraulic binders. Calcium carbonate may be beneficial.

4.3.1.4. Use with special cements

It has been reported that the effectiveness of calcium chloride in accelerating the strength gain of concrete containing pozzolans is proportional to the amount of cement in the mixture [39]. Various effects may be produced when calcium chloride is used as an admixture in concrete containing shrinkage compensating cement. While the use of 1-2% calcium chloride by weight of cement is common practice, its use will generally result in reduced expansion and increased drying shrinkage [67]. When used with Type M shrinkage compensating cement, calcium chloride will generally act as a retarder, owing to its effectiveness in retarding the hydration of aluminates. With all three types of shrinkage compensating cements, calcium chloride accelerates early strength development, owing probably to accelerated ettringite formation [68]. The limited, conflicting data on the effect of acceleration on the performance of concrete containing shrinkage compensating or self-stressing cements suggest that the concrete proposed for use should be evaluated with the accelerating admixture.

Calcium chloride should not be used with calcium aluminate cement since it retards the hydration of the aluminates. Similarly, calcium chloride and potassium carbonate retard the setting time of regulated set cement and lower the early strength developed by the calcium fluoroaluminate component. However, strengths after one day are improved by these additions. There are insufficient data on the use of accelerators with Portland blast-furnace slag cement or other blended cements to justify any conclusions on the effects of their combination in concrete. The use of admixtures with these cements should be preceded by an exploratory testing programme.

4.3.2. Retarders

Retarding admixtures are used to prolong and control the setting time of concrete by decreasing the initial rate of reaction between the cement and water without influencing the ultimate strength. These materials are particularly useful for offsetting the effects of high temperature, which decreases setting times, or to prevent complications when unavoidable delays occur between mixing and placing.

These admixtures can be divided into several categories on the basis of their chemical composition: (1) lignosulphonic acids and their salts, (2) hydroxycarboxylic acids and their salts, (3) sugars and their derivatives, and (4) inorganic salts. Lignosulphonate based admixtures are prepared from wastes from the pulp and paper industry, and studies indicate that most of the retarding properties of these admixtures may be due to compounds that belong to categories 2 or 3. Some inorganic salts (e.g. borates, phosphates, zinc and lead salts) can act as retarders but are not used commercially.

Retarders slow down the rate of early hydration of C_3S by extending the length of the dormant period. Thus the setting time of Portland cement, as measured by penetration tests, is extended. However, hydration in subsequent stages may be more rapid, so that development of strength need not be much slower than in an unretarded paste, if retardation is not excessive. The extension of the dormant period is proportional to the amount of retarding admixture that is used, and when the amount exceeds a critical point, C_3S hydration will never proceed beyond the dormant period and the cement paste will never set. Thus it is important to avoid overloading a concrete with a retarding admixture.

Retarders also tend to retard the hydration of C_3A and related aluminate phases. Thus hydroxycarboxylic acids, such as citric acid or gluconic acid, are used to control the handling time of regulated set cement. The interaction between C_3A and a retarder is complex. There is evidence that the initial reactions between C_3A and water may be accelerated even though the overall hydration is retarded, and the admixture may be incorporated into the hydration products of C_3A during their formation. Observations of C_3A–admixture interactions have led to the following conclusions [44]:

(a)   The effectiveness of a retarder depends on the C_3A content of the cement. (More retarder is removed from solution during the formation of the hydration products of C_3A, so that less is available to retard C_3S hydration.)

(b) The effectiveness of a retarder is increased if its addition to the fresh concrete is delayed for a few minutes. (Less retarder is removed from solution, since some hydration products have already formed before the admixture is added.)

(c) A retarder may cause abnormal setting problems with particular cements: both early stiffening and abnormal retardation of setting have been observed. Cements from the same mill may behave differently with a water reducing admixture, depending on the SO_3 content of the cement. (The aluminate/sulphate ratio may be thrown out of balance when the retarders change the rate of hydration of C_3A.)

(d) An admixture may extend the setting time but not the time during which the concrete can be handled and placed. (The early reactions of C_3A affect workability, and accelerating the early hydration may promote slump loss.)

(e) A retarding admixture may influence the behaviour of an expansive cement. (The rate of formation of ettringite is changed.)

In addition, the content of alkali oxides in a cement may determine the effectiveness of a retarder by causing the breakdown of the admixture. Because retarders are so sensitive to the composition of the cement, an admixture should be evaluated with the cement that is to be used on the job.

As mentioned above, the retarding power of an admixture increases when its addition to concrete is delayed a few minutes after the first addition of water. Retardation is increased rapidly up to about 10 min delay (depending on admixture type and dose) and then slowly decreases (Fig. 16). After a delay of 2–4 h the admixture will no longer retard the setting time. Consequently, an additional dose of a

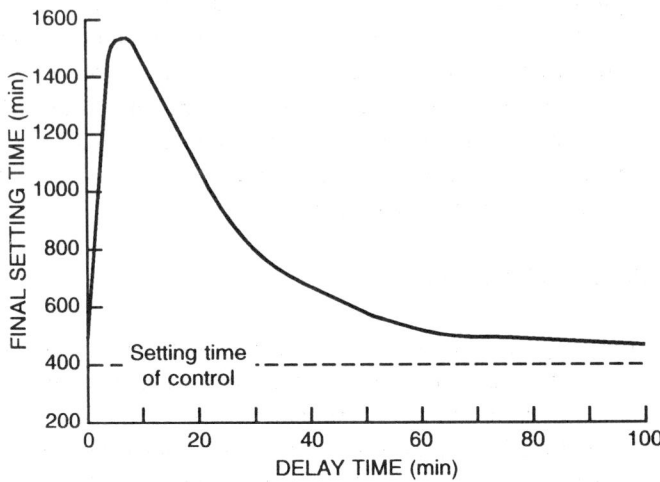

FIG. 16. Effect of delayed addition of retarding admixture on its retarding power [12].

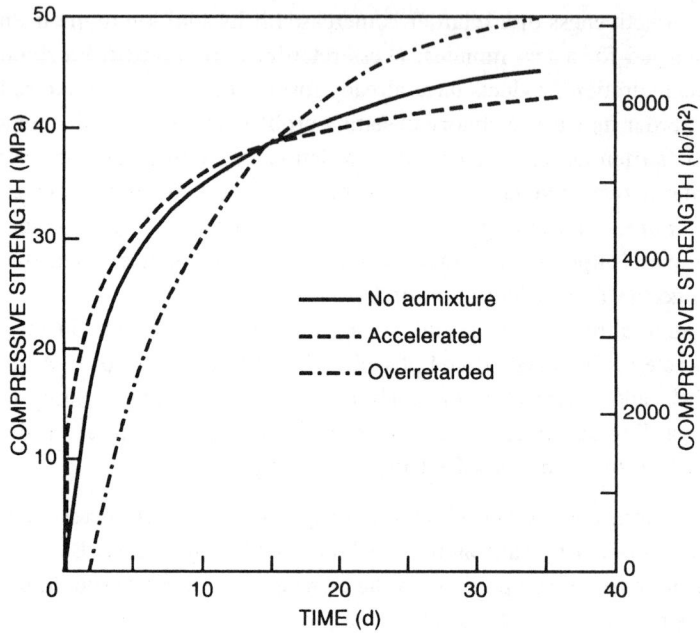

FIG. 17. Effect of set modifying admixtures on strength development of concrete [12].

water reducing admixture can be used during retempering of fresh concrete without unduly prolonging setting times. Whenever a retarding admixture is used, some reduction in the one day strength of the concrete should be anticipated. Within seven days, the strength should approach that of an unretarded concrete, unless an overdose has been used (Fig. 17). Retarding admixtures have been reported to increase ultimate compressive strength and, to a lesser extent, flexural strength. Although set controlling admixtures are reported to increase drying shrinkage and creep, the effects depend on changes in mix design, time of hydration and time of drying or loading. Laboratory work indicates that admixtures may increase the rate of drying shrinkage and creep but not the final values [12].

4.3.3. Water reducing admixtures

Water reducing admixtures lower the amount of water required to attain a desired slump, which means an effective lowering of the w/c ratio, with a consequent general improvement in strength, watertightness (impermeability) and durability. Alternatively, the desired slump may be achieved without changing the w/c ratio by lowering the cement content. This may be done for economic or for technical reasons (e.g. to lower the heat of hydration of the cement).

According to ASTM C494, an admixture can be classified as water reducing if it reduces water requirements by 5%. Most conventional admixtures achieve water reductions of 5–10% at normal dosages, but some newer admixtures, called 'super-plasticizers', can cause reductions of 15–30%.

The materials that are generally available for use as water reducing admixtures fall into five general classes:

(1) Lignosulphonic acids and their salts.
(2) Modifications and derivatives of lignosulphonic acids and their salts.
(3) Hydroxylated carboxylic acids and their salts.
(4) Modifications and derivatives of hydroxylated carboxylic acids and their salts.
(5) Other materials, which include:
   (i) Inorganic materials, such as zinc salts, borates, phosphates and chlorides;
   (ii) Amines and their derivatives;
   (iii) Carbohydrates, polysaccharides and sugar acids;
   (iv) Certain polymeric compounds, such as cellulose ethers, melamine derivatives, naphthalene derivatives, silicones and sulphonated hydrocarbons.

These admixtures can be used alone or in combination with other organic or inorganic materials. Admixtures of all five classes are available as either powders or liquids.

Superplasticizers, the latest in the wide range of admixtures that have changed traditional concrete practice, have two main functions: (1) the provision of exceptionally high slump, while retaining cohesion; and (2) the reduction of water content at normal workability, with resultant gain in strength. The beneficial influence of

TABLE XIV. EFFECT OF DOSAGE OF SUPERPLASTICIZER ON WATER REDUCTION [69]

Dosage[a]	Cement content (kg/m^3)	Slump (mm)	w/c	Water reduction (%)
—	309	50	0.49	—
0.64	309	64	0.43	15
1.73	326	64	0.35	25
2.30	329	108	0.35	24
2.82	332	132	0.35	24

[a] wt% of solids in cement.

TABLE XV. COMPOSITION AND APPLICATION OF SUPERPLASTICIZERS

Brand name	Form	Active ingredient	Effects	Dosage	Specific gravity (g/mL)	Shelf life
PSP-R	Liquid	Modified naphthalene sulphonate	Water reduction 25%, retardation 1–3 h	5–20 mL/kg	1.205	1 a
PSP-N	Liquid	Modified naphthalene sulphonate	Water reduction up to 25%	5–20 mL/kg	1.200	1 a
Mighty 150	Liquid	Sulphoaryl alkylene	Water reduction 15–20%	0.6–1.2 wt%	1.20	—
Mighty 150-RD2	Liquid	Sulphoaryl alkylene + carboxylic acid salt	Water reduction 15–20%, set retardation	0.5–0.7 wt%	1.20	—
Mulcoplast CF	Liquid	Sulphonated polymer	Water reduction up to 25%	7.55–20 mL/kg	1.07–1.09	>1 a
Sikament	Liquid	Anionic dispersant	Water reduction up to 30%	7.8–23.4 mL/kg	1.2	>1 a
Melment-L10	Liquid	Sulphonated melamine–formaldehyde	Water reduction up to 30%	1.5–3.0 wt%	1.11	>1 a
Melment-F10	Solid	Sulphonated melamine–formaldehyde	Water reduction up to 30%	0.3–0.6 wt%	—	>1 a

superplasticizers on concrete properties has been proved by laboratory research and site application, and growing use has confirmed their freedom from side effects. Admixtures of this kind give exceptional flowability or high strength to concrete, beyond the limits of plasticizers. Their value extends to mortars, grouts, rendering and pumping, particularly where the concrete would otherwise be too stiff for compaction.

Superplasticizers act by physical absorption, not chemical reaction with the cement or aggregates. They do not modify the mature properties of the concrete, but reduce friction within the mix so that the concrete flows freely for a considerable period.

These materials are effective because undesirable side effects, air entrainment and set retardation, are absent or very much reduced. Thus, high amounts can be used (Table XIV), typically exceeding 1 wt% of active ingredient in cement, whereas large quantities of conventional water reducers cannot be used. Superplasticized concretes can be air entrained for freeze–thaw resistance, although the characteristics of the air void system are different from those of conventional concrete with the same air content.

The chief advantages of superplasticizers may be summarized as follows:

(a) Little compaction of the concrete is required, with savings in handling and placing.
(b) w/c ratios are reduced, as are the cement content and heat of hydration.
(c) High early strengths are attainable.
(d) Setting times are unaltered.
(e) Very little air is entrained, even at high dosages.
(f) As a result of lower permeability, there is better resistance to weathering and chemical attack.
(g) The admixtures are compatible with all types of Portland cement.

Superplasticizers were first developed in Japan and the Federal Republic of Germany and have recently gained widespread acceptance. Table XV gives information on their properties [70].

4.4. MINERAL ADMIXTURES

In Europe, large quantities of blended Portland cements containing pozzolanic and cementitious additives are made. For various reasons, there is more interest in the USA and Canada in using industrial by-products as mineral admixtures in concrete rather than as components of blended cements. However, the underlying mechanisms by which the combination of pozzolanic and cementitious materials with Portland cement influences the engineering behaviour of the product are not significantly affected by the way in which these by-products are incorporated into concrete.

TABLE XVI. CLASSIFICATION, COMPOSITION AND PARTICLE CHARACTERISTICS OF MINERAL ADMIXTURES FOR CEMENT [78]

Classification	Chemical and mineralogical composition	Particle characteristics
Cementitious and pozzolanic		
Granulated blast-furnace slag	Mostly silicate glass containing mainly calcium, magnesium, aluminium and silica. Crystalline compounds of melilite group may be present in small quantities.	Unprocessed material is of sand size and contains 10–15% moisture. Before use it is dried and ground to particles of less than 45 μm (usually about 500 m^2/kg Blaine). Particles have rough texture.
High calcium fly ash	Mostly silicate glass containing mainly calcium, magnesium, aluminium and alkalis. The small quantity of crystalline matter present generally consists of quartz and C_3A; free lime and periclase may be present; CS and C_4A_3S may be present in the case of high sulphur coals. Unburnt carbon is usually less than 2%.	Powder corresponding to 10–15% particles larger than 45 μm (usually 300–400 m^2/kg Blaine). Most particles are solid spheres of less than 20 μm diameter. Particle surface is generally smooth but not as clean as low calcium fly ash.

Highly active pozzolans

Condensed silica fume	Consists essentially of pure silica in non-crystalline form.	Extremely fine powder consisting of solid spheres of 0.1 μm average diameter (about 20 m^2/g surface area by nitrogen adsorption).
Rice husk ash (Mehta–Pitt process)	Consists essentially of pure silica in non-crystalline form.	Particles are generally smaller than 45 μm but they are highly cellular (about 60 m^2/g surface area by nitrogen adsorption).

Normal pozzolans

Low calcium fly ash	Mostly silicate glass containing aluminium, iron and alkalis. The small quantity of crystalline matter present consists generally of quartz, mullite, sillimanite, haematite and magnetite.	Powder corresponding to 10–15% particles larger than 45 μm (usually 200–300 m^2/kg Blaine). Most particles are solid spheres of average 20 μm diameter. Cenospheres and plerospheres may be present.
Natural materials	Besides aluminosilicate glass, natural pozzolans contain quartz, feldspar and mica.	Particles are ground to mostly smaller than 45 μm and have rough texture.

Weak pozzolans

Slowly cooled blast-furnace slag, bottom ash, boiler slag, field burnt rice husk ash	Consists essentially of crystalline silicate minerals with only a small amount of non-crystalline matter.	The materials must be pulverized to very fine particle size in order to develop some pozzolanic activity. Ground particles are rough in texture.

Since natural pozzolans and industrial by-products generally cost substantially less than Portland cement, the exploitation of the pozzolanic or the pozzolanic and cementitious properties of mineral admixtures when used as a partial replacement for cement can yield large economic benefits. Possible technological benefits from the use of mineral admixtures in concrete include enhancement of impermeability and chemical durability, improved resistance to thermal cracking and increase in final strength. In addition to the excellent report on pozzolans and pozzolanic cements by Lea [11], several extensive reviews have been published on the properties of cements and concretes involving natural pozzolans, fly ash and blast furnace slags [71–77].

Granulated blast-furnace slag and low calcium fly ash have long been used as Portland cement additives or as mineral admixtures in concrete. For historical reasons, the science and technology associated with these materials developed along separate lines. However, with the advent of many additional by-products, such as high calcium fly ash, rice husk ash, condensed silica fume, and granulated non-ferrous slags, there is a need to treat the entire subject as a unified discipline.

For the most part the natural pozzolans are materials of volcanic origin, but include certain diatomaceous earths. The artificial pozzolans are mainly products obtained by the heat treatment of natural materials such as clays, shales, certain siliceous rocks and pulverized fuel ash (fly ash).

A new classification of mineral admixtures has been suggested (Table XVI) based on their mineralogical composition and particle characteristics which influence the cement–admixture interaction [78].

4.4.1. Natural pozzolans

Except for diatomaceous earths, all natural pozzolans are derived from volcanic rocks and minerals. On the basis of the principal lime reactive constituent present, the natural pozzolans can be classified into volcanic glasses, volcanic tuffs, calcined clays and shales, and diatomaceous earths.

Deposits of natural pozzolans are located throughout the world, and important material characteristics from some well known pozzolan deposits are described below. Analytical oxide contents of typical samples of natural pozzolans are shown in Table XVII.

4.4.1.1. Volcanic glasses

Bacoli pozzolan of Italy, Santorin earth of Greece and Shirasu pozzolan of Japan are examples of the pozzolanic materials which derive their lime reactivity characteristics mainly from unaltered aluminosilicate glass. As a rule, small amounts of non-reactive crystalline minerals such as quartz, feldspar and mica are embedded in the glassy matrix. Mineralogical analysis of a specimen of Shirasu pozzolan showed 95% glass content, with quartz and anorthite as principal crystalline impuri-

50

TABLE XVII. OXIDE CONTENTS IN NATURAL POZZOLANS [11, 71, 73, 77]

Pozzolan	SiO_2	Al_2O_3	Fe_2O_3	CaO	MgO	Na_2O	K_2O
Volcanic glasses							
Bacoli pozzolan (Italy)	53.1	18.2	4.3	9.0	1.2	3.1	7.6
Santorin earth (Greece)	65.1	14.5	5.5	3.0	1.1	2.6	3.9
Shirasu pozzolan (Japan)	69.3	14.6	1.0	2.6	0.7	3.0	2.4
Volcanic tuffs							
Segni–Latium (Italy)	45.5	19.6	9.9	9.3	4.5	0.9	6.4
Rhenish trass (Germany)	52.1	18.3	5.8	4.9	1.2	1.5	5.1
Bavarian trass (Germany)	62.4	16.5	4.4	3.4	0.9	1.9	2.1
Higashi–Matusyma (Japan)	71.8	11.5	1.1	1.1	0.5	1.5	2.6
Calcined clays and shales							
Handelage (Germany)	42.2	16.1	7.0	21.8	1.9	0.30	1.0
Diatomites							
Diatomaceous earth (USA)	86.0	2.3	1.8	—	0.6	0.4	—

ties [73]. The vitreous matrix of the Bacoli pozzolan shows inclusions of leucite, feldspar and augite [71]. Quartz and feldspar (anorthite and labradorite) are the main crystalline impurities found by X ray diffraction analysis of typical samples of Santorin earth.

4.4.1.2. Volcanic tuffs

The pozzolans of Segni–Latium (Italy) and trass of the Rheinland and Bavaria (Germany) are typical volcanic tuffs. The zeolite tuffs, with their compact texture, are fairly strong, possessing compressive strengths of the order of 10–30 MPa [72]. The principal zeolite minerals are reported to be phillipsite and herschelite. After the compact mass is ground to fine particles, the zeolite minerals show considerable reactivity with lime and develop cementitious characteristics similar to those of the pozzolans containing volcanic glass.

4.4.1.3. Calcined clays and shales

Volcanic glasses and tuffs do not require any heat treatment to enhance their pozzolanic property. However, clays and shales cannot produce appreciable hardening with lime unless the crystal structures of the clay minerals are destroyed by heat treatment in an industrial furnace. Calcination temperatures of 600–900°C in oil, gas or coal fired rotary kilns are considered adequate for this purpose. The pozzolanic activity of the calcinated product is mainly due to the formation of an amorphous or disordered aluminosilicate structure. Surkhi, a pozzolan made in India by pulverizing fired clay bricks, belongs to this category. The heat treatment of clays and shales which contain large amounts of quartz and feldspar would yield good pozzolans.

Production of pozzolans from calcination of clays is not now favoured, owing to the large energy requirements of the process. Compared with pozzolanic industrial by-products such as fly ash that are generally available in large quantities, the calcined clays and shales have become uneconomical.

4.4.1.4. Diatomites

This group of pozzolans is characterized by materials of organogen origin. Diatomite is a hydrated amorphous silica, composed of skeletal shells from the many species of microscopic aquatic algae. The largest known deposit is in California, USA; other large deposits occur in Algeria, Canada, Denmark and Germany. Diatomites are highly reactive to lime, but their skeletal microstructure accounts for a high water requirement, which decreases the strength and durability of concrete containing this pozzolan. Furthermore, diatomite deposits, such as the Moler deposit in Denmark, generally contain large amounts of clay and therefore calcination is required to enhance the pozzolanic reactivity.

4.4.2. Artificial pozzolans

Ashes from the combustion of coal and some crop residues, volatilized silica from certain metallurgical operations, and granulated slag from the ferrous and non-ferrous metals industries are the major industrial by-products which are suitable for mineral admixtures in Portland cement concrete. Their production and properties are described in this section.

Industrialized countries such as France, Germany, Japan, the former USSR, the UK and the USA are among the largest producers of fly ash, volatilized silica and granulated blast-furnace slag [79]. In addition, China and India have the potential for making large amounts of rice husk ash.

The present rate of production of fly ash in the world is estimated to be about 180 million tonnes per year. In 1980, the USA produced 48.3 million tonnes of fly ash, and 2.67 million tonnes were used in cement and concrete products. Sixty per cent of the annual production of volatilized silica products came from the USA, the former USSR and Norway. Large quantities of blast-furnace slag are available in many countries; however, the former USSR, France and Germany are major producers of granulated slag that possesses pozzolanic and cementitious properties. In the USA, at present only 1.6 million tonnes are being granulated from the 26 million tonnes of blast-furnace slag produced each year. The granulation and use of steel slag and non-ferrous slags as mineral admixtures in concrete are still in the stage of research and development.

4.4.2.1. Fuel ash (fly ash)

During combustion of powdered coal in modern power plants, as coal passes through the high temperature zone in the furnace, volatile matter and carbon are burned off, whereas most of the mineral impurities such as clays, quartz and feldspar are fused. The fused matter is quickly transported to lower temperature zones where it solidifies as spherical particles. Some of the mineral matter agglomerates to form bottom ash, but most of it flies out with the flue gas stream and is called fly ash. This ash is subsequently removed from the gas by electrostatic precipitators and is the type normally used for cementing and grouting purposes. Fly ash contains a glassy silica phase which reacts with lime produced by the hydration of Portland cement to produce calcium silicate hydrate gel. The low heat of hydration of cements which incorporate fly ash is an advantage in full scale operations, where it reduces the curing temperature [80–82].

4.4.2.2. Rice husk ash

Rice husks, also called rice hulls, are the shells remaining from the dehusking of paddy rice. Since they are bulky, the husks present an enormous disposal problem for centralized rice mills. Each tonne of paddy rice produces about 200 kg of husks which, on combustion, yield about 40 kg of ash. The ash from open field burning or from uncontrolled combustion in industrial furnaces which use husks as fuel consists mainly of crystalline silica minerals such as cristobalite and tridymite, and must be ground into very fine particles to develop pozzolanic properties. On the other hand, the ash produced at low temperatures in a process developed by Mehta and Pitt contains silica in a cellular, high surface area and non-crystalline form (50–60 m^2/g), and is therefore highly pozzolanic [83].

4.4.2.3. Volatilized silica

Volatilized silica, sometimes known simply as silica fume or condensed silica fume, is generated by electric arc furnaces as a by-product of the production of metallic silicon or ferrosilicon alloys. In the reduction of quartz to silicon at temperatures of up to 2000°C, gaseous SiO is produced. It is transported to lower temperature zones, where it oxidizes on contact with air and condenses into spheres consisting of non-crystalline silica. The material, which is extremely fine, is removed by filtering the outgoing gases in a bag filter. Like rice husk ash, condensed silica fume is highly pozzolanic because it consists essentially of non-crystalline silica with a very high surface area (20–23 m^2/g).

4.4.2.4. Blast-furnace slag

In the production of pig-iron in a blast-furnace, if the slag is cooled slowly in air, its chemical components are usually present in the form of crystalline melilites (Ca_2AS–C_2MS_2 solid solution), which do not react with water at ordinary temperatures. If ground to very fine particles, the material will be weakly cementitious and pozzolanic. However, when the liquid slag is rapidly quenched from a high temperature (1400–1500°C), either by water or by water and air, most of the lime, magnesia, silica and alumina can be held in a non-crystalline or glassy state. The water quenched product is called granulated slag, owing to the sand size particles, whereas the slag quenched by water and air is in the form of pellets and is called pelletized slag. Normally the former contains more glass; however, when ground to 400–600 m^2/kg Blaine surface area, both products can develop satisfactory cementitious and pozzolanic properties.

4.4.2.5. Other slags

Steel slags are a by-product of the conversion of pig-iron to steel. Basic oxygen furnaces and open hearth and electric arc furnaces are used for this purpose. Steel slag is similar in chemical composition to iron blast-furnace slag except that it usually contains a large proportion of iron oxide and smaller amounts of silica and alumina. Steel slag cooled slowly in air is virtually inert. However, reactive slag for use as a cementitious pozzolanic admixture for concrete can be produced by water granulation. In the production of metallic copper, nickel and lead, large amounts of slag are produced by the various smelting furnaces. All these slags contain unusually high amounts of iron oxide (40–60%). Copper and nickel slags are also characterized by a low lime content. On water quenching and fine grinding, they show pozzolanic behaviour. On the other hand, lead slags contain 10–20% CaO, and should therefore develop cementitious properties.

4.4.2.6. Properties of pozzolanic cement concretes

The substitution of a pozzolan for Portland cement reduces the strength obtained at the earlier ages, though the ultimate strength attained may be increased.

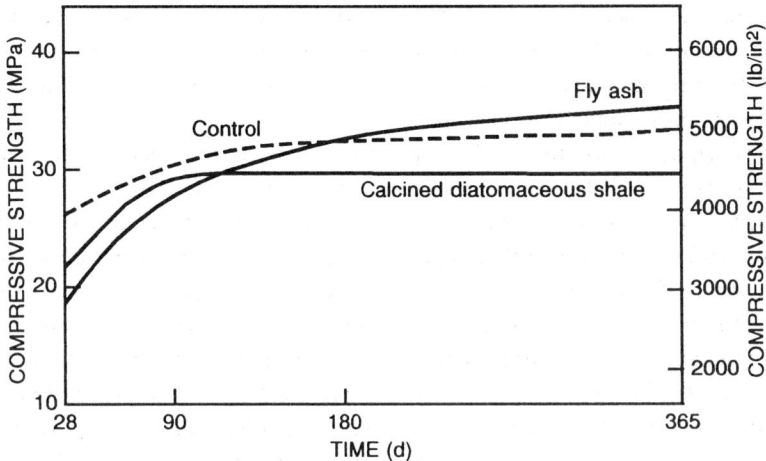

FIG. 18. *Effect of a pozzolan on the compressive strength of concrete (30 wt% replacement of cement) [84].*

TABLE XVIII. COMPRESSIVE STRENGTH[a] OF POZZOLANIC CEMENT

Pozzolan	Portland cement A			Portland cement B		
	28 d	180 d	1 a	28 d	180 d	1 a
Burnt clay	68	82	93	61	69	77
Burnt shale	74	88	95	64	74	80
Rhenish trass	69	75	80	58	66	71
Naples pozzolan	60	71	79	52	63	70
Ground sand	45	52	57	45	49	50
Strength of 100% Portland cement (kg/cm^2)	252	405	432	395	544	568

[a] Expressed as a percentage of the strength of 100% Portland cement at the same age.

This change is illustrated in Fig. 18, where the compressive strength of concrete made with pozzolanic cements containing 30 wt% of pozzolans is shown [84].

Table XVIII compares several pozzolans and shows that the proportionate loss in strength at ages up to a year with a pozzolan is greater with a rapid than with a slower hardening Portland cement [4].

Concretes containing pozzolanic cements usually have a lower slump than those made with 100% Portland cement. The plasticity of the concrete and its freedom from segregation and bleeding, at any slump, are usually much improved, and this is particularly notable in lean mixes.

Besides improving the workability of harsh mixes, pozzolans lower the total heat of hydration and also the rate of liberation of heat, because the pozzolanic reaction is quite slow. The speed of the pozzolanic reaction is comparable to that of C_2S hydration and the result of adding a pozzolan is to effectively raise the C_2S content of the cement. Thus, a Type I cement can be used with a pozzolanic admixture as a substitute for Type IV in mass concrete. As a result, Type IV cements are now seldom manufactured. The use of pozzolans also improves the impermeability and durability of hardened concrete. Expansion resulting from the alkali–aggregate reaction is reduced with a pozzolan and a Type I cement can be used with a pozzolan that is low in alumina to replace Type V cement in concretes exposed to sulphate attack. It is believed that the reduction of calcium hydroxide in the paste is an important factor in improving chemical durability. The amounts of additions depend on the particular application, but may be as high as 35 wt% of the cement.

Because the continuing formation of hydrates fills the pores and because there is no free lime which could be leached out, partial replacement of Portland cement by a pozzolan reduces the permeability of concrete [9–12].

4.5. MISCELLANEOUS ADMIXTURES

Recently, the polymer latexes, such as styrene–butadiene rubber, polyacrylic ester, poly(vinylidene chloride–vinyl chloride), poly(ethylene–vinyl acetate) and polyvinyl acetate latexes, have been widely employed as admixtures. In Japan, several standards for the quality and testing of latex type cement modifiers and latex modified mortars were issued as Japanese Industrial Standards (JISs) (Table XIX).

Latex modification of cement is governed by both the processes of cement hydration and polymer film formation in the binder phase. Cement hydration generally precedes the process of polymer formation [85]. A matrix phase which consists of cement gel and polymer film may be formed as a binder [86–88]. The process of polymer film formation on the cement hydrates is represented in Fig. 19. It is generally considered that hardened cement paste has mainly an agglomerated structure of calcium silicate hydrates and calcium hydroxide bound together by the weaker van der Waals force; therefore, microcracks occur easily in the paste under stress. This structure gives poor tensile strength and fracture toughness to ordinary cement mortar and concrete. By contrast, in the latex modified mortar and concrete the microcracks appear to be bridged by the polymer films or membranes, which prevents crack propagation and, simultaneously, a strong cement hydrate–aggregate bond develops [86].

TABLE XIX. JAPANESE INDUSTRIAL STANDARDS FOR LATEX TYPE CEMENT MODIFIERS AND FOR TESTING OF LATEX MODIFIED MORTARS

JIS A 1171	Making test sample of polymer modified mortar in the laboratory
JIS A 1172	Test for strength of polymer modified mortar
JIS A 1173	Test for slump of polymer modified mortar
JIS A 1174	Test for unit weight and air content (gravimetric) of fresh polymer modified mortar
JIS A 6203	Polymer dispersions for cement modifiers

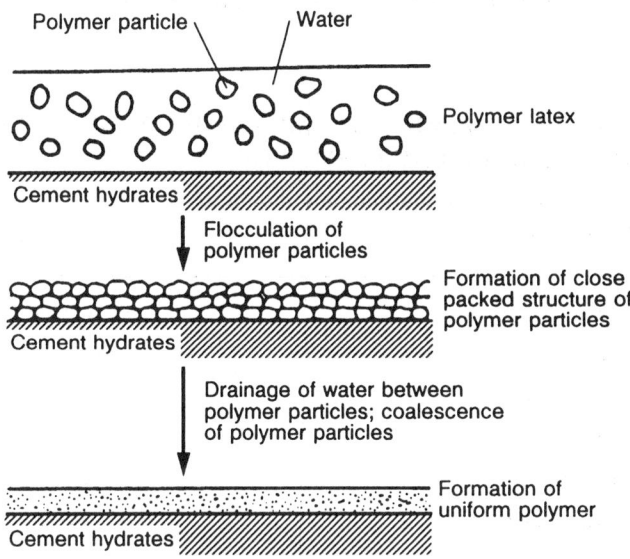

FIG. 19. *Simplified model of the process of polymer film formation on cement hydrates.*

The latter effect increases with an increase in the polymer content, and leads to increased tensile strength and fracture toughness. However, entrainment of excess polymer and air entrainment cause discontinuities in the formed monolithic network structure, whose strength is thereby reduced. The sealing effect due to the polymer films or membranes formed in the solidified waste form improves the properties of the waste form, considerably increases waterproofing or watertightness, and improves resistance to moisture transmission, chemical resistance and freeze–thaw durability. Such effects not only improve the leachability but also the long term durability of the waste form in its disposal environment by increasing resistance to aggressive waters.

5. CHEMICAL ASPECTS OF WASTE CEMENTATION

5.1. INTRODUCTION

Solidification of wet solid wastes with hydraulic cement, either with or without admixtures, has been practised for many years. Water in the waste reacts chemically with the cement to form hydrated silicate and aluminate compounds, which normally contribute towards the setting and hardening of the cement mixture. However, the wastes produced by nuclear activities are very diverse and under certain circumstances affect the rate of hydration of cement and/or reduce the quality of the product. Optimal formulations should consider each waste individually because of possible interactions between the constituents of the waste and the cement. Table XX summarizes the chemical compatibility of various types of wastes and hydraulic cements.

TABLE XX. CHEMICAL COMPATIBILITY OF WASTES WITH HYDRAULIC CEMENTS [89]

Waste type	Waste compatibility
Ion exchange resins	Poor–good[a]
Sludges	Good
Boric acid wastes	Poor–good[a]
Sulphate wastes	Fair
Nitrate wastes	Good
Phosphate wastes	Good
Carbonate wastes	Good
Detergent solutions	Poor–good[b]
Complexing agent wastes	Poor
Oils, organic liquids	Poor–good[c]
Acidic wastes	Poor–good[d]
Alkaline wastes	Good
Filter cartridges	Good

[a] Good with admixtures.
[b] Good with antifoaming agents.
[c] Good with emulsifying agents.
[d] Good with neutralizing agents.

5.2. CHEMICAL PRETREATMENT OF WASTE

Chemical pretreatments are applied to waste streams to reduce or eliminate potentially adverse waste–cement interactions, or to effectively encapsulate the waste homogeneously in a cement matrix. This section identifies waste streams which are considered 'problem wastes' with respect to solidification with cement. Problem wastes are defined as those wastes which cannot be solidified, are solidified with difficulty, have poor volumetric efficiencies or produce an unsatisfactory product. The majority of these wastes are liquid and wet solid waste streams containing dissolved or suspended chemical components that interfere with cement hydration reactions.

5.2.1. Boric acid wastes

A principal waste material generated by PWRs consists of hot (approximately 77°C) aqueous concentrates of boric acid, typically 5–12 wt%. The aqueous concentrates are maintained at elevated temperatures prior to incorporation into cement to facilitate pumping of waste as a slurry, ensure a more homogeneous waste–cement mix and accelerate the setting of the mix. Direct solidification of this waste with Portland cement is complicated by the retarding effects of both the acidity of the waste and the specific action of the borate ion on the setting of cement [26, 27].

Boric acid wastes can be neutralized with sodium hydroxide solution; however, this introduces more water into the system and does not reduce the level of soluble borate. Solidification can be accomplished by the use of excess caustic and cement, but this results in poor waste loading efficiency.

The precipitation of borates by the addition of calcium ions to the waste before immobilization prevents the encapsulation of unhydrated cement particles by formation of calcium metaborate ($CaO \cdot B_2O_3 \cdot mH_2O$). The calcium salts thus formed have limited solubility over a broad temperature range, and thereby control the borate ion concentration before the addition of cement as well as provide the alkaline pH required for setting of the cement mixture. The insoluble borate salts also contribute to the plasticity of the mix, facilitating reduction of the cement without forming a permanent bleed liquid.

Accelerators such as calcium chloride and sodium hydroxide are not essential for cement solidification of borated solutions, but their addition in controlled amounts ensures more rapid exothermic development in bulk mixtures, which reduces the waste processing time per batch. The normally elevated temperature of the waste is actually an advantage in promoting more rapid set. There is some speculation that at least part of the calcium borate present during solidification may be converted to more stable zeolitic structures that could be integrated closely or bonded with the hydrated cement. The zeolite mineral datolite contains a borate unit and has the formula $2SiO_2 \cdot 2CaO \cdot B_2O_3 \cdot mH_2O$, which is similar to certain complex cement hydrates. The exothermic temperatures, sometimes reaching 90°C, would favour the development of such compounds during bulk solidification.

Without pretreatment of boric acid waste, a maximum of 5 wt% boric acid can be incorporated into Portland cement. After pretreatment, as much as 50 wt% borate salts can be incorporated into the same bulk solidifications.

An alternative to chemical pretreatment of the waste is masonry (high lime) cement. A 1:1 mixture of slaked lime and Portland cement can incorporate approximately 15 wt% boric acid [27]. Details of the masonry cement–boric acid waste system are given in Section 3.

5.2.2. Sodium sulphate wastes

Waste from BWRs contains 8–25 wt% soluble sodium sulphate salts. Sodium sulphate reacts exothermally with cement hydration products, such as tricalcium aluminate ($3CaO \cdot Al_2O_3$), and can promote an undesirable 'flash' set. The addition of a molar excess of lime before adding the cement converts the sodium sulphate to calcium sulphate (gypsum), with a moderating effect on the evolution of heat. The benefits of lime addition for improved reaction control are still offset by excessive exotherms for large bulk solidifications. While the gypsum that is formed slows the reactions, the by-product is sodium hydroxide, which accelerates setting.

$$Na_2SO_4 + Ca(OH)_2 \longrightarrow CaSO_4 + 2NaOH$$

Depending upon the initial concentration in the matrix, sulphates further react both with the free calcium hydroxide to form calcium sulphate and with the hydrated calcium aluminates to form calcium sulphoaluminates.

Positive control can also be achieved by the addition of 0.5–1.0% boric acid to the cement mixture. Reductions in peak exothermic temperatures of 12–22°C are achieved by this means, without loss of product strength.

5.2.3. Ion exchange resin wastes

The effect of resin beads on cement has been studied extensively [4, 90, 91]. The water demand by the resin, its increase in volume with subsequent swelling of the hardened cement and the ion exchange action which continues within the cement matrix have been documented.

Resins containing complexing agents such as borates and organic acids (from decontamination) are difficult to solidify directly with cement and first require special treatment. With an excess of acids, the pH has to be adjusted and the exchange reaction with calcium ions should be complete. Failure to pretreat the cement waste may result in incomplete hydration and setting, and give a matrix that is less water resistant and subject to bead swelling to a point where the product deteriorates.

The cement content should be sufficient to separate individual beads and the w/c ratio should be carefully controlled to maintain product hardness and strength.

A mechanism which may improve the product quality is to minimize the hydrated lime phase in blast-furnace cement or Portland cement, since in the case of cation exchangers it has been found that the degradation of the cemented waste form is associated with the expansive growth of hydrated lime crystals after setting has occurred.

5.2.4. Particulate sludges

Particles with neutral chemical activity such as Solka-Floc and diatomaceous earth do not interfere with the hydration reactions of cement and may actually improve the integrity of the product. Finely powdered ion exchange materials, when stabilized in the calcium form by the addition of appropriate amounts of lime or sodium carbonate, do not affect cement setting reactions (see Section 5.2.3).

5.2.5. Iron hydroxide flocs

Iron hydroxide flocs are produced by treatment of effluents to remove actinides from liquid waste streams. Studies of simulated waste have shown that these flocs retard the hydration of blast-furnace slag cements and produce a product with high shrinkage, leading to disruption of the matrix. Special cements are required, as the use of ordinary Portland cement results in an unacceptably high exotherm. Gypsum cements are being considered for these wastes in the UK, but work is at an early stage of development. High alumina cements may also be considered.

An alternative option which is currently under consideration in the UK is a pretreatment stage before encapsulation. Two principal methods are being tested:

(a) The addition of calcium hydroxide, which produces a chemical reaction with the floc to yield calcium ferrite;
(b) Floc dispersal flocculation using proprietary agents to modify the properties of the floc.

5.2.6. Reactive metals

Metals such as magnesium and aluminium may be important components of radioactive wastes. Large quantities of magnesium waste result from reprocessing Magnox reactor fuel elements, whereas aluminium is a component of cartridge filters and miscellaneous solid wastes.

Solidification of these metals in cement requires careful consideration of the amount of metal that can be safely encapsulated and of the encapsulation matrix, because of the corrosive reaction which occurs at high pH in cement. Such reactions may lead to expansion and eventually to disruption of the matrix and to the production of hydrogen. These reactions take place at the surface of the metal. By limiting the surface area of the metal encapsulated and by using low permeability matrices such as blast-furnace slag cements, these deleterious corrosion reactions can be controlled to produce acceptable waste forms.

6. CEMENT SOLIDIFICATION PROCESSES

6.1. INTRODUCTION

This section outlines the design principles and plant requirements of the two types of cementation systems which use in-container and in-line mixing and summarizes their advantages, disadvantages and development status. The application of these systems to in situ immobilization is discussed.

6.1.1. Requirements for a solidification system

The basic requirements for solidification systems are: simplicity of operation, reliability, ease of maintenance, minimum radiation exposure of personnel during operation and maintenance, reasonable cost, and a systematic procedure for providing reasonable assurance that the solidified product is consistent.

The choice of process will also be influenced by factors such as the nature of the waste, its specific activity, the processing rate required, the properties to be achieved in the immobilized waste and its disposal route. The economics of the waste conditioning process will be influenced by these interacting factors.

6.1.2. Process control programmes

In all types of solidification systems, the sensitivity of the cementation process to changes in waste composition, proportions of the mix and variations in any pretreatment chemistry must be known, since this affects the degree of measurement and control required in feed systems for cement, waste and other components of the mix. A systematic process control programme is required to provide reasonable assurance that the solidified product will meet established criteria for solidified waste. Such a programme consists of two parts. The first part is a set of bounding values for system and waste parameters within which satisfactory solidification can be expected to occur with a high degree of confidence [92]. This can be illustrated by the 'composition' diagram in Fig. 20. To satisfactorily solidify with cement the

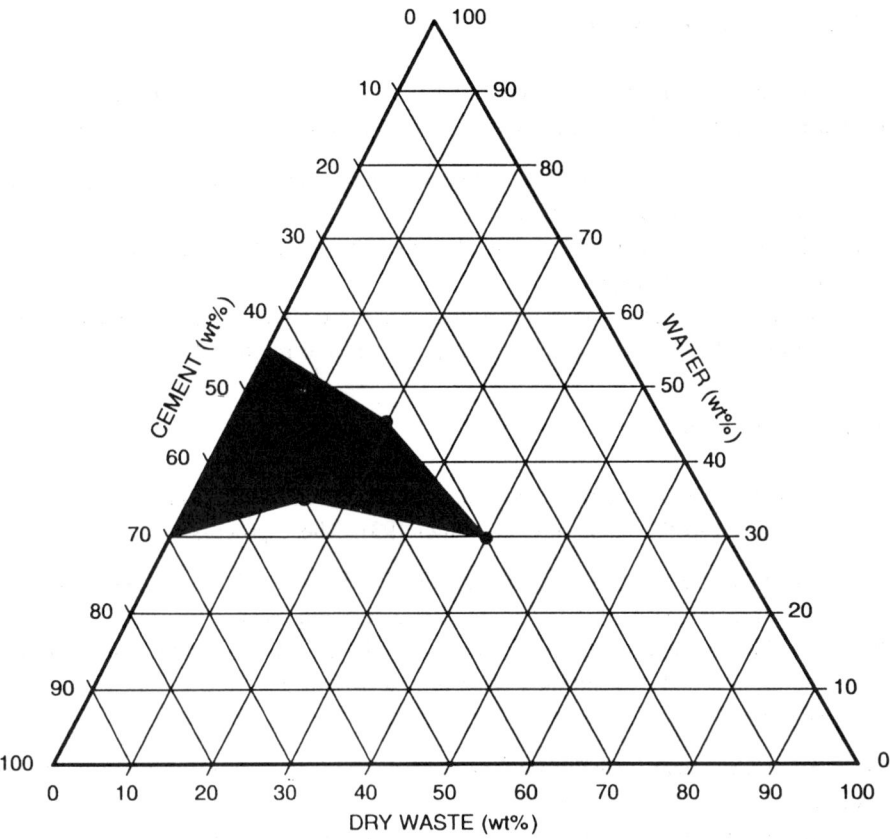

FIG. 20. Process control programme composition diagram.

waste represented in the diagram, the product must comprise between 0 and 40 wt% dry waste, between 30 and 55 wt% water, and between 30 and 70 wt% cement.

The second part of a process control programme is a systematic procedure using appropriate controls and instrumentation to demonstrate that the solidification system can operate within the specified boundaries. As part of the programme, periodic checks of the product should be made to verify that the control parameters are providing the required properties in the solidified waste.

6.2. MIXING PROCESSES

Processes for mixing the waste with cement can be divided into two categories: in-container and in-line.

6.2.1. In-container mixing

In-container mixing processes involve mixing the waste and cement inside a disposable container [93]. The mixing techniques used include the following:

(a) Tumbling or rolling a sealed container with (or in some systems without) a disposable mixing weight inside;
(b) Mixing with a disposable mixer that stirs the contents of the container and is left in the container;
(c) Mixing with a reusable element that stirs the contents of the container and is removed before the container is capped and the mixture hardens.

6.2.2. In-line mixing

In-line mixing includes all processes where the waste and binder are mixed and then transferred to the disposable container [93]. These processes include:

(a) Batch mixing where the waste and binder are mixed in a process vessel and then transferred to the disposable container as a batch;
(b) Continuous mixing where the waste and binder are continuously metered into the mixer and the mixture is continuously transferred to the disposable container.

Admixtures can be incorporated into the cement mixture as required at the appropriate location in the system.

In general, provisions must be made for flushing the system and for processing the material that is flushed out.

6.3. IN-CONTAINER CEMENTATION PROCESSES

6.3.1. In-drum tumbling process

An economical process developed in the USA uses in-drum 'tumble' mixing of cement with radioactive liquids and slurries [89]. Figure 21 is a schematic of this process. The procedure uses 55 gal (208 L) drums of the closed top design, with a 4 in (10 cm) screw type cap at the centre of the lid. Cement and a mixing weight are placed in a drum through the top opening, and the drum is then transferred to a processing unit where the waste is added. The drum is then capped and tumbled end over end to mix the waste and cement. The drum is reopened, the void is filled, and the drum is capped and tumbled again. The drum is then stored while the mixture solidifies. The cement is measured by weight and the waste is transferred to the drum with positive displacement metering pumps.

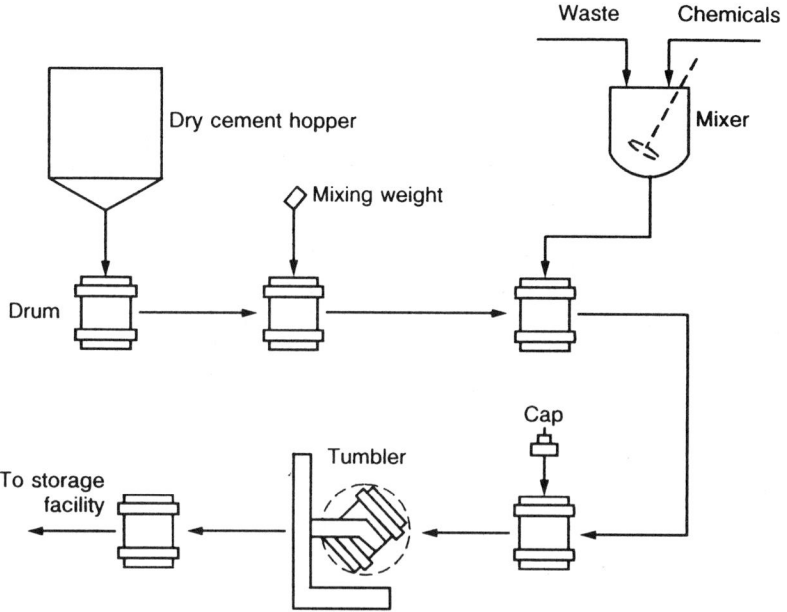

FIG. 21. In-drum tumble process [89].

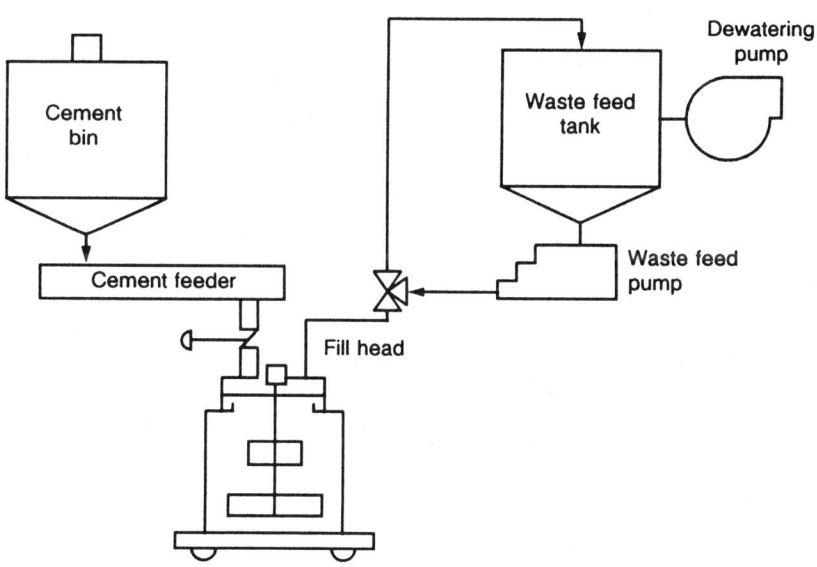

FIG. 22. Flow diagram for an in-container cement solidification system.

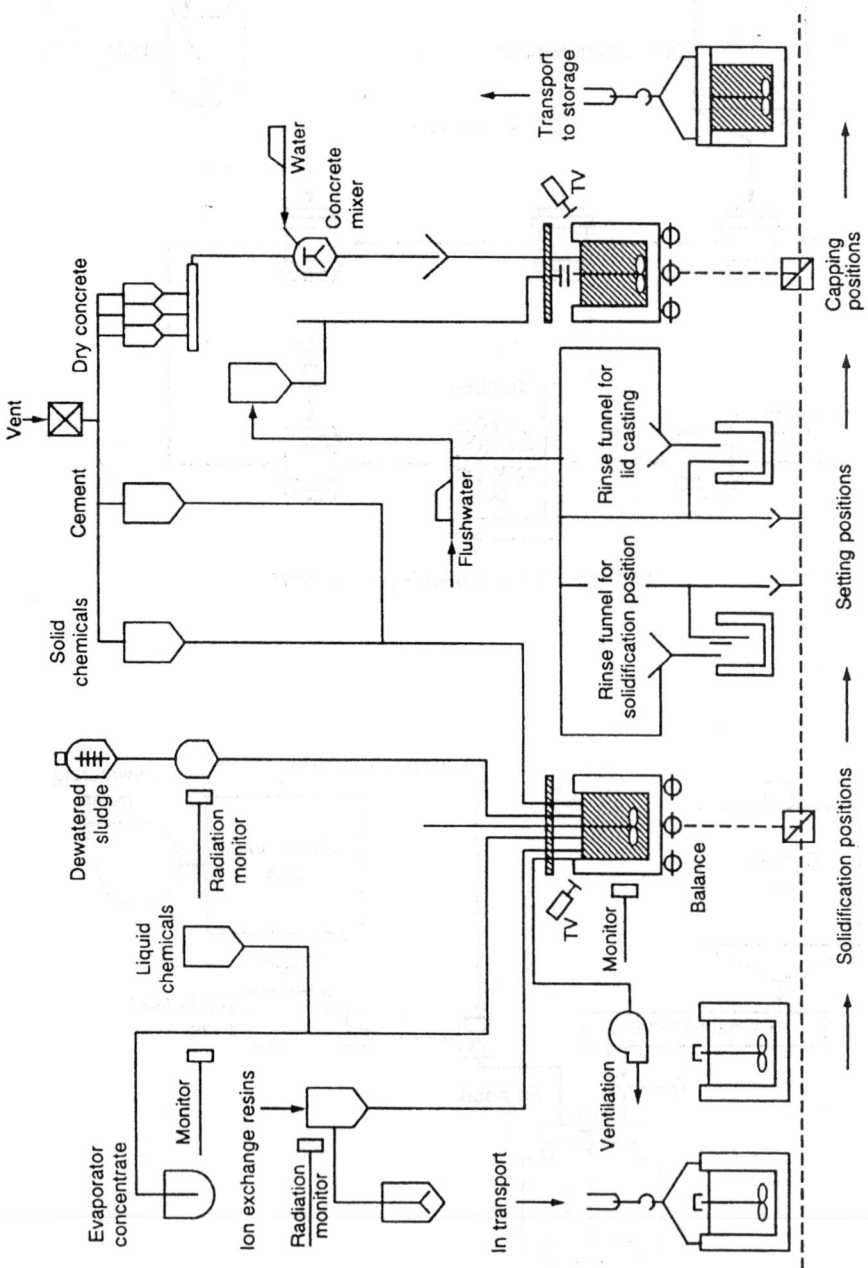

FIG. 23. ABB ATOM waste solidification system [94].

6.3.2. Disposable mixer processes

Disposable mixing blades are usually placed in the container in an inactive area and the container–mixing blade assembly enters the plant as one unit. The drive to the mixer is engaged within the active part of the plant, either by raising the drum to the mixer drive, or by lowering the mixer drive onto the container. At the end of the mixing the drive is disengaged and the mixing blade remains in the container as the cement hardens. This procedure prevents any secondary wastes arising from cleaning the mixing blades.

A simplified schematic of an in-container batch mixing process is shown in Fig. 22. This system can be used to immobilize liquid wastes, ion exchange resins, or dried wastes from incinerators or other volume reduction systems. Polymers can be used as the solidification agent. Container sizes may vary from 200 L to 5.7 m^3. There are provisions to obtain waste samples from the system for analysis and quality control.

The waste is mixed with the cement in the drum or shipping liner, using a disposable agitator, and allowed to solidify. The container is then remotely capped and prepared for storage or shipment.

A solidification system developed by ABB ATOM AB in Sweden uses cement to solidify waste in concrete containers. The capacity of these containers is approximately 1 m^3 [94]. In the ABB ATOM system, filter sludges and powdered resins are dewatered to 20–30 wt% dry content. Granular resins are conveyed by an air lift to a dosage tank and the batch of settled resin is then flushed into the waste container by adding a specified amount of water.

Before dry cement is added, certain admixtures are added to improve the properties of the final product. A complexing agent is added to nullify the effects of boric acid wastes, which retard the setting of the cement.

After solidification, a non-radioactive concrete cap is cast on top of the waste–cement mixture to fill the container void space. The waste package is transported from the solidification process to storage by a vehicle that is programmable and remotely controlled. Figure 23 is a flow diagram illustrating the ABB ATOM waste solidification system.

6.3.3. Non-disposable mixer processes

Non-disposable mixing blades are temporarily engaged to the mixer. This allows an efficient mixing system for obtaining homogeneity of the waste. However, careful control of the process is required to ensure that the blade is removed before the cement begins to set. During operation of the process, the mixer blades require cleaning daily. This can be done by operating the mixer in a special drum containing water and sand. Mixing blades also can be cleaned with water spray systems.

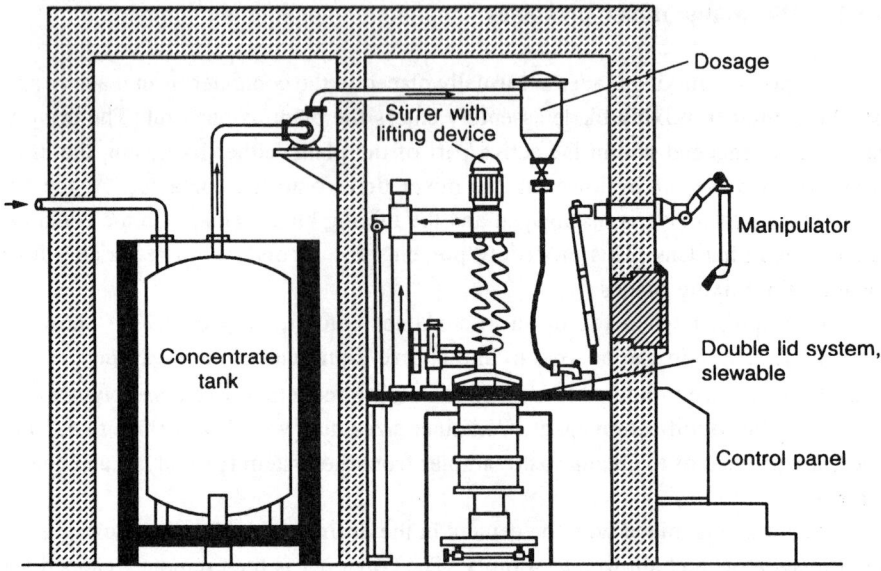

FIG. 24. System for cementation of radioactive concentrates, resins and sludges at the Karlsruhe Nuclear Research Centre.

In a concept developed by the Karlsruhe Nuclear Research Centre, solidification is carried out in a shielded concrete cell. Drums which are prefilled with cement from beneath are introduced into the cell by means of a double lid system. The liquid concentrates are transferred from a dosage vessel within the box to the drums by means of a vacuum pump. Concentrates and cement are then mixed with a planetary stirrer which is retracted before the cement mix begins to harden. Figure 24 is a drawing of the in-container solidification system.

6.3.4. Order of mixing

The two alternatives are to mix waste into a predispensed quantity of cement or to dispense cement into a predispensed quantity of waste. The former technique has an advantage in that the required amount of cement is added to the drum in a non-radioactive area before it enters the radioactive area of the plant. However, the addition of liquid wastes to dry cement powder makes mixing difficult; the design of the mixing paddle and the rotation speed must be carefully controlled in order to obtain a homogeneous mix without balling of solids. Adding cement powder to liquid wastes avoids these difficulties but must be done in the radioactive area. In both cases it is important to obtain a good seal between the top of the drum and the mixing–feed system to prevent the spread of radioactivity.

6.3.5. Advantages and disadvantages

The advantages and disadvantages associated with in-container mixing are summarized below [95]:

Advantages

— The system is simple, with all operations conducted in a single container.
— There are no secondary wastes, except for reusable mixing blades.
— Only limited quantities of waste are present in the system. Any difficulties can be corrected rapidly and are restricted to a single container.

Disadvantages

— The drum cannot be filled completely, since an ullage space is necessary to allow mixing to take place.
— Sampling of the cement is inconvenient, so the process relies on good control.
— Cleaning of the mixing blade is required when reusable blade systems are used.

6.4. IN-LINE CEMENTATION PROCESSES

6.4.1. High shear batch processes

A process of in-line high shear batch mixing is used in the USA for immobilizing liquid and wet solid wastes with cement. Other wastes, such as spent filter cartridges or contaminated articles, may be placed directly into drums for encapsulation.

The major components of the cement solidification subsystem include the high shear radwaste mixer, the waste dispensing system, the flushwater recycling system, the cement storage and feed system, and the container handling system. At the start of the waste solidification process, liquid wastes enter the dispensing vessel and are mixed by the agitator. The waste feed is then discharged into the high shear mixing vessel together with a proportionate amount of Portland cement from the cement storage and feed system. The mixer imparts high shear to the waste–cement mixture, creating a homogeneously mixed paste in which the cement particles are rapidly hydrated. After several minutes, the mixture is automatically discharged into waste disposal containers, capped and placed in storage. This system is illustrated in Fig. 25 [96].

An advanced batch process incorporates a highly reliable mixer that is widely used in civil engineering and in the chemical industry and that has been adapted to the nuclear industry. The basic principle is to mix radioactive waste and cement

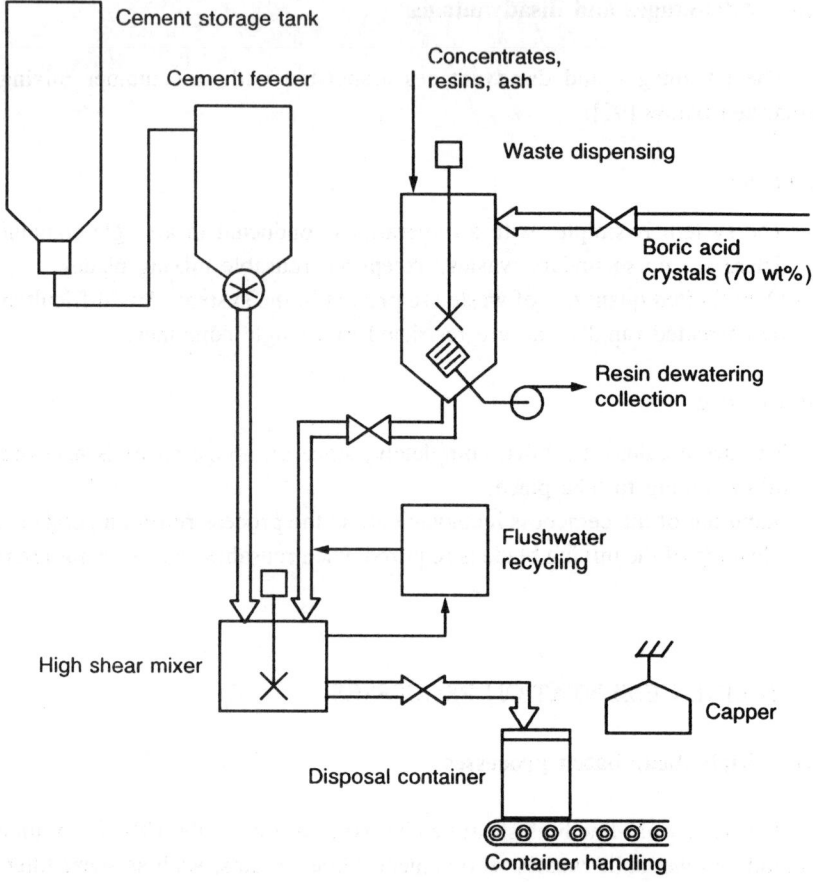

FIG. 25. Westinghouse high shear in-line batch cement solidification unit.

inside a batch mixer, whose capacity is adjusted to the required throughput and container type.

The radioactive waste is transferred into the batch mixer, which can be equipped with weighing probes, giving a precise control of the amount of waste. Chemical adjustments or pretreatments may be made in the waste storage tank or in the mixer. The proportions of the cement mix and additives are precisely controlled by the weighing probes as the cement and additives are fed into the mixer. Mixing is completed within a few minutes and the cement grout is poured into the container through a discharge valve. A schematic of this batch system is shown in Fig. 26. This system, developed by SGN (Société générale pour les techniques nouvelles) is used for cementation of alpha bearing waste at the Commissariat à l'énergie atomique facility in Valduc, France.

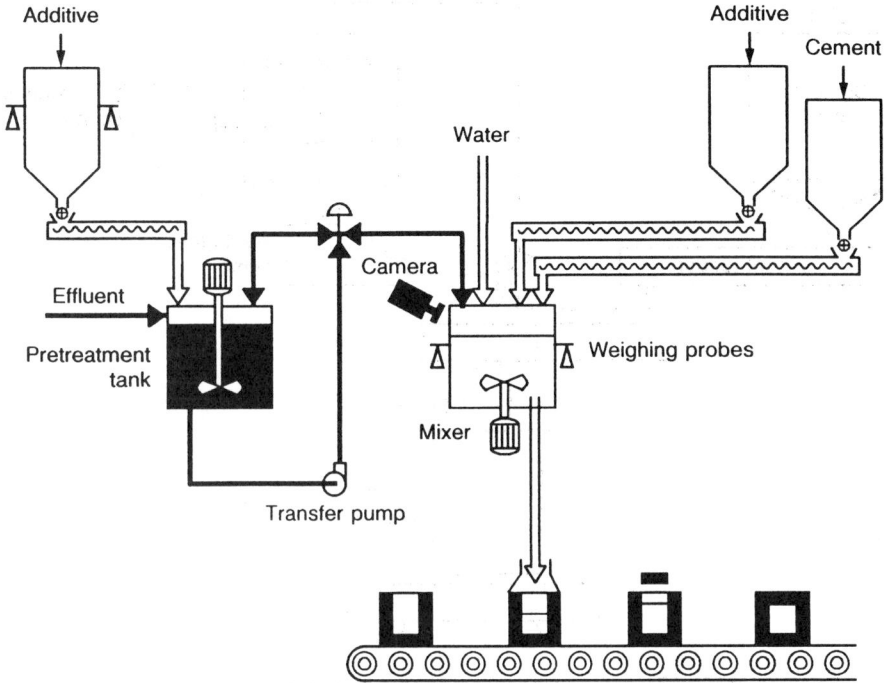

FIG. 26. *Schematic of an improved in-line batch cementation process.*

6.4.2. Continuous processes

An in-line immobilization process can be used to solidify wet and dry wastes from incinerators or volume reduction systems. The waste and solidification agents are metered into a continuous mixer, where they are mixed. Container sizes may vary from 55 gal (208 L) drums to 200 ft^3 (5.66 m^3) liners. Provisions are made to obtain samples of the waste before solidification to ensure that nothing will interfere with processing. The mixture is pumped into a container where it solidifies. The container is remotely capped and stored. Provisions are made for dewatering resin and filter sludges in the waste feed tank. Process flow diagrams for typical in-line continuous processes are shown in Figs 27 and 28 [96].

The advantages and disadvantages associated with continuous mixing processes are summarized below [95]:

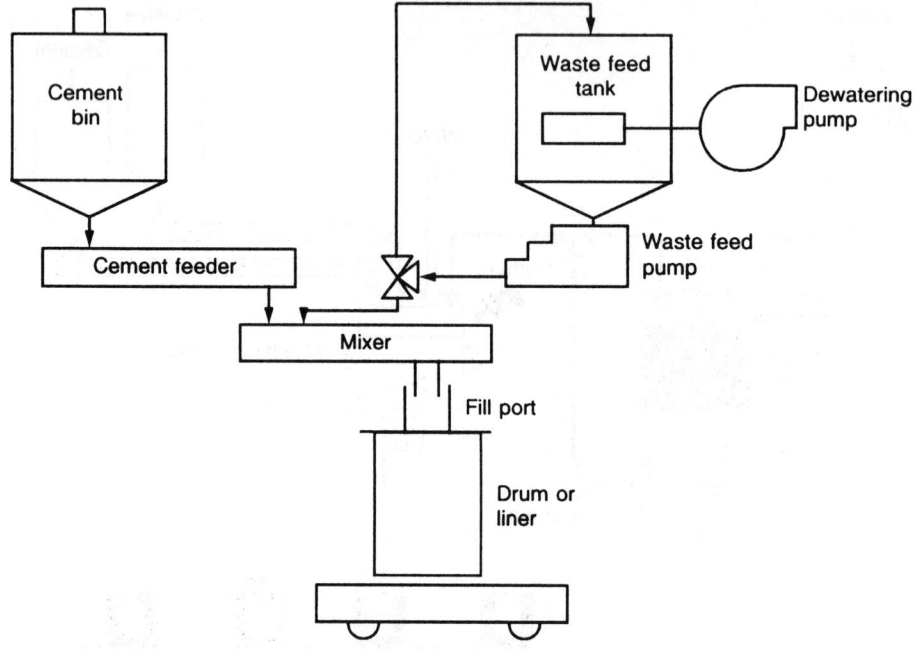

FIG. 27. *In-line continuous cement solidification system.*

Advantages

    — Size of the mixer: small hold-up; small plant, decreasing cost.
    — Short residence time, allowing a high throughput.
    — High quality mixing.
    — Easy sampling.
    — Optimization of drum filling.
    — Can be fitted with every type of container.

Disadvantages

    — There are strict requirements for process control.
    — The clean-out procedure must be introduced.
    — The process must operate continuously in order to avoid repetition of shutdown/startup procedures.
    — There may be problems connected with inhomogeneity of the waste.
    — Maintenance is more important than in batch mixing.
    — Blockages may possibly occur (in feeding equipment, etc.).

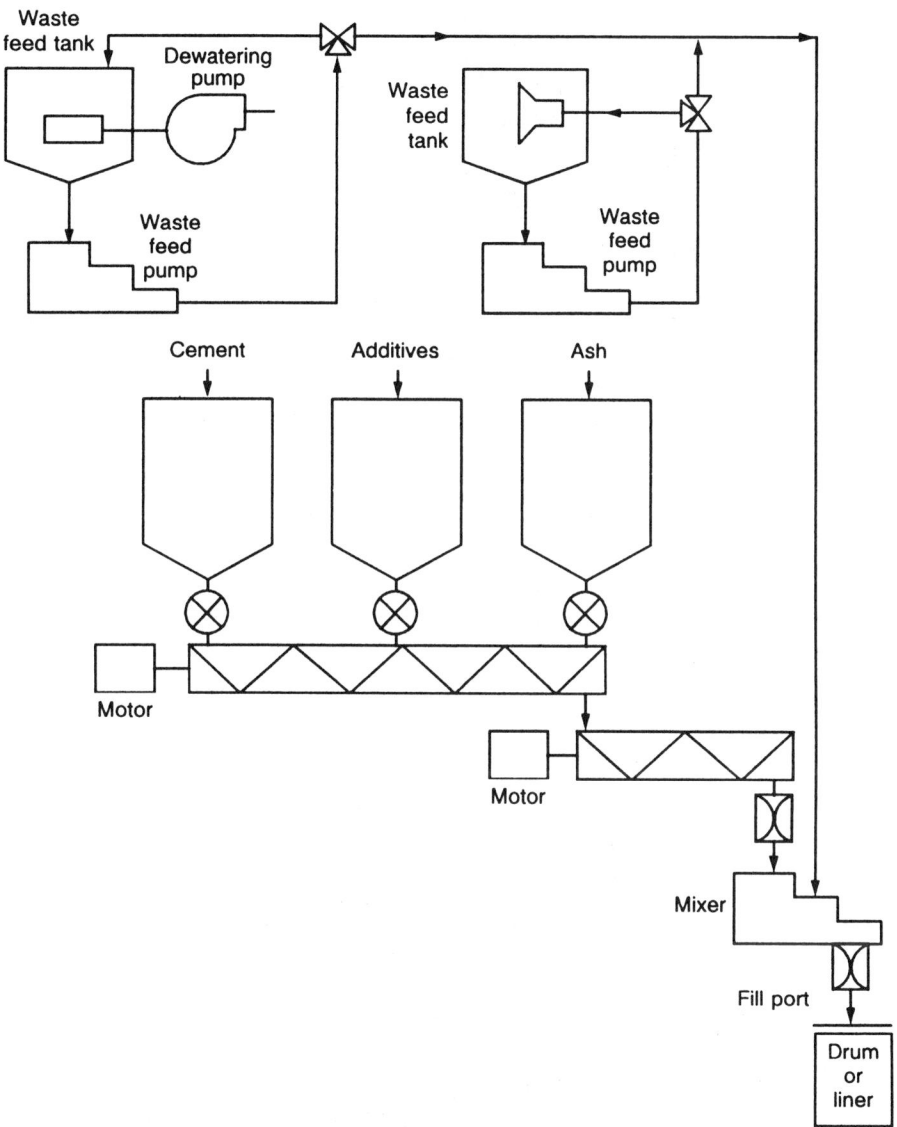

FIG. 28. In-line continuous cement solidification system.

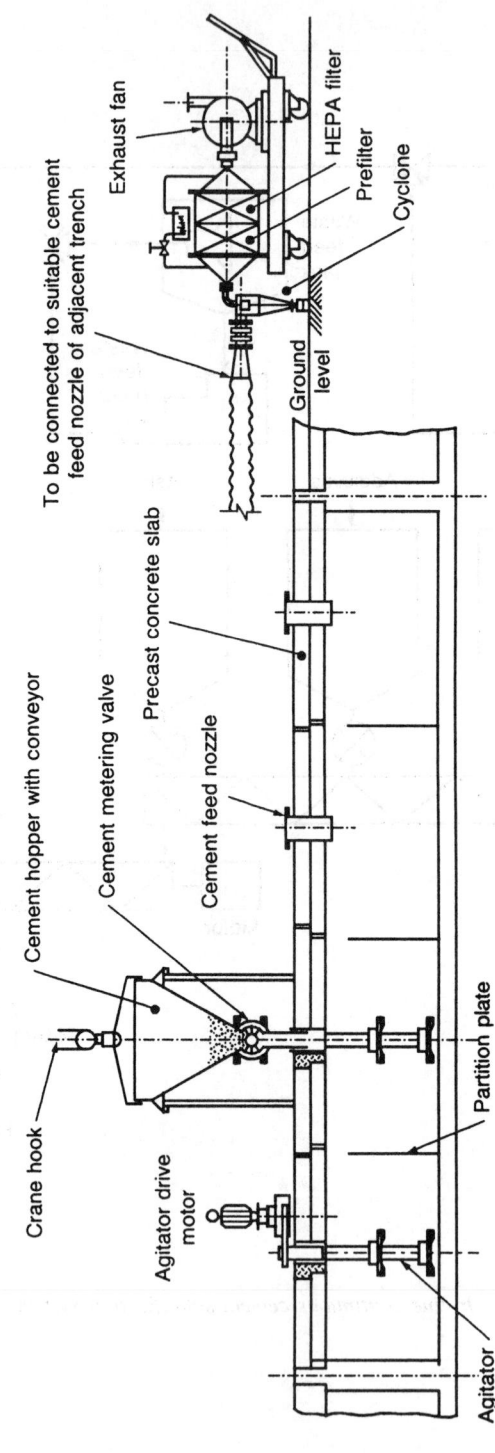

FIG. 29. *In situ solidification of sludge with cement.*

The various types of in-container and in-line mixing processes are summarized below:

In-container mixing processes

— Rolling
— Rotary paddles
  Reusable
  Disposable
— Tumbling.

In-line mixing processes (all mixers are continuous except for the high speed, high shear, low pressure batch mixer)

— Extruders
— High shear kneading and screw auger
— High speed, high shear, low pressure batch mixer
— Positive displacement pumps
— Screw augers
— Static mixers.

6.5. IN SITU IMMOBILIZATION PROCESSES

6.5.1. In-container processes

The in situ mixing techniques are site specific and can be practised where the disposal site is near the place of waste generation or waste treatment. This method, shown in Fig. 29, is in use for solidification of chemical sludge generated in a low level liquid radioactive waste treatment plant.

The disposal trench consists of a series of below ground reinforced concrete containers. Their size and number are determined by the quantity of waste to be solidified. Each container is fitted with a mixing blade and provided with nozzles for waste inlet, cement, admixtures and ventilation. The top of the container is closed with a concrete shield plug sufficiently thick to provide adequate shielding for personnel.

Waste is fed into the containers using a shielded tanker pumping the waste directly from a nearby waste treatment plant through underground piping.

After thorough homogenization of the sludge with any necessary admixtures such as vermiculite, cement is added under continuous agitation of the sludge with the installed mixing blade. The amount of cement added is metered by a rotating valve volume control. After the process is complete, the detachable equipment (cement hopper, mixer motor, ventilation equipment) is moved to a new container for the next operation.

The top of the cemented waste is capped with a cement grout after the matrix has hardened for several days. The nozzles are seal welded and the top of the trench is sealed with a concrete mix, followed by a waterproofing treatment.

The advantages and disadvantages of the in situ solidification processes are summarized below:

Advantages

— Waste is fixed directly into the disposal ground, eliminating the need for packaging, transport and further handling.
— The ratios of waste, additives and cement can be varied according to the results of laboratory studies.
— No maintenance is necessary on contaminated equipment since the mixer is disposable and is used for only a few hours at the time of fixation.
— Exposure to the operator is negligible.
— There is no secondary waste from washdown or decontamination.
— Processes are easy to control if the agitator fails.

Disadvantages

— The processes are site specific.
— Advance construction of disposal containers and mixers is required.
— Frequent handling of material is required for feeding cement, moving the mixing motor from one container to another, and connecting and disconnecting the ventilation equipment.
— The cost of the disposable type agitator is high.

6.5.2. Continuous process

In this process radioactive waste and cement mix are continuously fed into a mixer, the formulation being adapted to maintain the flowability of the cement paste to allow its continuous discharge or pumping towards the disposal trench or concrete underground pool. A drawing of this process is shown in Fig. 30 [97, 98].

Apart from the requirements common to every continuous cementation process there are two main additional requirements:

— On-site disposal must be available.
— Discharging the cement waste by pumps and a network of pipes, or any other system, may require modified cement formulations and flushing procedures.

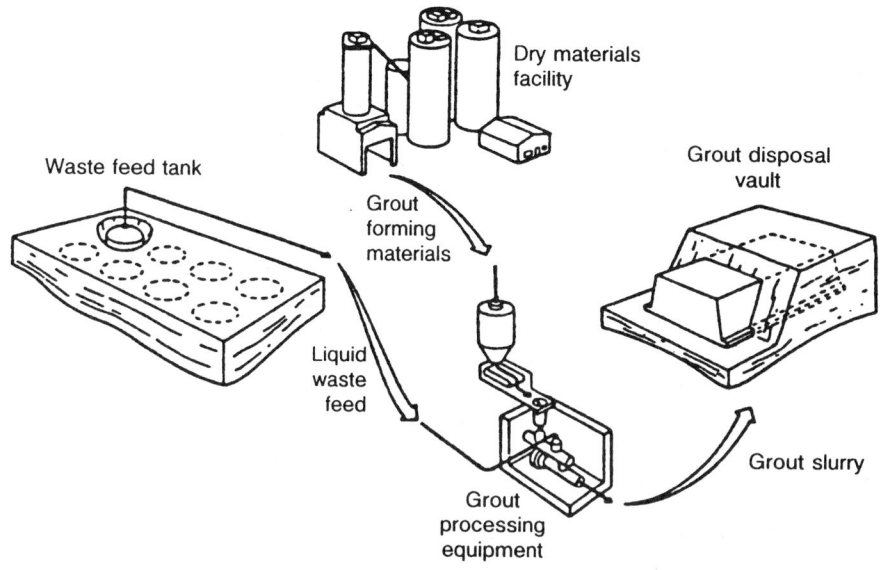

FIG. 30. *Continuous in situ solidification process.*

6.6. MOBILE CEMENTATION UNIT

A mobile solidification unit was designed specifically to facilitate in-container solidification of various radioactive wastes, including evaporator bottoms, resin beads, powdered resin, filter sludges, filter precoat backwash, resin regeneration chemical wastes (sodium sulphate), decontamination solutions and oil.

The waste is immobilized using readily available Portland type cements in conjunction with hydrated lime and proprietary additives. All conditioning and solidification of the waste occurs in a disposable container. Depending on the characteristics of the waste, some of the required conditioning chemicals may be loaded into the disposable container before the waste. Preinstalled mixer blades permit continuous agitation until a thick cement paste has formed that will set to a hard, homogeneous, water free matrix.

Specific chemical additives are used to control pH, increase the matrix strength, reduce leachability, increase the ratio of waste to end product, reduce the heat generated, prevent 'bleed' liquid, and control and adjust cement setting times. A balanced chemistry uses the waste components themselves to enhance solidification. This procedure makes possible a predictable end product.

The disposable container is prepared by installing temperature detector leads and a level measurement tube which connects to the fill head. Then in the process area the disposable container is placed into a shipping cask or shield (if required).

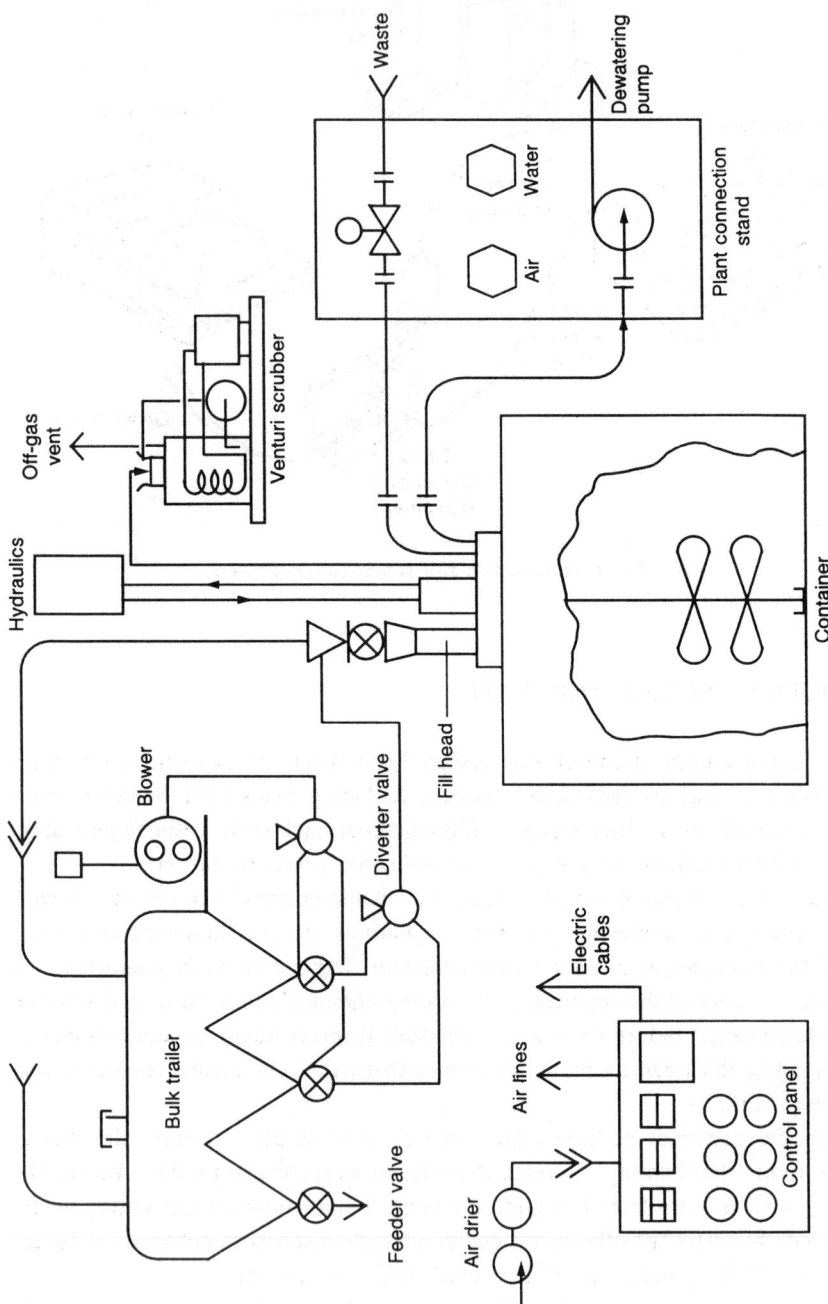

FIG. 31. Mobile cement solidification unit.

The fill head–skirt assembly is installed on the container. The hydraulic motor shaft engages with the disposable mixer blade assembly to provide the agitation required. The dewatering leg is then connected, if required, for dewatering the waste slurry. A flow diagram for this process is shown in Fig. 31.

7. PROPERTIES OF IMMOBILIZED WASTE FORMS

7.1. INTRODUCTION

A large portion of the waste produced in the nuclear industry is either liquid or in a wet solid form (e.g. resins and filter sludges) and requires conditioning to achieve an acceptable solid, monolithic form for disposal.

The structural stability of the conditioned waste is a primary concern. A structurally stable waste form will generally maintain its physical dimensions under the expected disposal conditions, including such factors as the weight of the overburden and compaction equipment, the presence of moisture, microbial activity, internal factors such as radiation effects, and chemical changes.

The waste form must have properties that provide integrity during handling, storage, transportation and disposal, and the long term stability of the waste form in disposal environments.

Specific waste form property requirements are dependent on the type of disposal to which the waste form will be subjected. Near surface disposal, below ground vaults, above ground vaults, earth mounded bunkers and mined cavities may require different levels of waste form performance. Other factors that influence waste form property requirements depend on site characteristics such as surface water, groundwater, geology, ecology and socioeconomic factors.

In general, the following guidelines provide for stability of waste forms [99]:

(a) The waste should not contain free standing, drainable or corrosive liquids.
(b) The waste form should be resistant to degradation caused by radiation effects.
(c) The waste form should be resistant to biodegradation.
(d) The waste form should remain stable under the compressive loads inherent in the disposal environment.
(e) The waste form should remain stable if exposed to moisture or water after disposal.
(f) The as-generated waste should be compatible with the solidification media.

The following are some of the most critical properties that must be considered to ensure cement waste form durability:

(1) Chemical stability
 — Waste–solidification agent interaction
 — Leachability
 — Waste form–container interaction
 — Waste form–disposal environment interaction
(2) Mechanical stability
 — Compressive strength
(3) Thermal stability
 — Freeze–thaw resistance
(4) Radiation stability
 — Chemical and mechanical characteristics
 — Gas generation
(5) Microbial degradation.

7.2. CHEMICAL STABILITY

The concept of chemical stability for cementitious materials is complicated by the chemical reactions taking place between the solidification agent and waste components. A broad overview of phenomena relevant to the performance of conditioned waste forms follows.

7.2.1. Waste–solidification agent interactions

The potential problems encountered during the solidification of specific waste streams in cement have been discussed in Section 5 of this report. These reactions are typified by inhibition, retardation or acceleration of the hydration reactions. In less dramatic cases, chemical interactions between waste and cement occur without affecting the hydration reactions. These reactions proceed at a slow rate and eventually result in the deterioration of the waste form in storage or disposal.

The most commonly used technique for determining waste–solidification agent interactions is by total immersion of the test specimen in an aqueous medium for a prolonged period (e.g. 90 d). The aqueous medium can be distilled water, site specific groundwater, brine or sea water. This test is based on the premise that chemical reactions between waste and cement are accelerated by total immersion in water. During immersion the samples are examined for swelling, cracking, loss of compressive strength, etc.

80

7.2.2. Leachability

Leachability is the measure of radionuclides and other contaminants transported from the waste form to the environment by an aqueous medium. Leach tests are generally used for purposes such as:

— Comparison of solidification agent–waste combinations,
— Evaluation of additives and admixtures,
— Determination of release rates,
— Determination of leaching mechanisms.

A variety of established leaching tests are available and the use of an appropriate test method depends to a large extent on the disposal scenario. The better known leach tests are: static [100], semidynamic [101, 102], accelerated [103], and flow-through [104].

Leaching from cement based waste forms is controlled by several release mechanisms (diffusion, adsorption plus diffusion and solubility limits), depending on the element being considered [105]. Transition metals, such as Co, Ni and Fe, and transuranic elements generally leach at very low rates because the high pH of cement pore water requires that these elements be in forms that have extremely low solubilities. Release rates of Cs are usually high, controlled essentially by the porosity and tortuosity of the waste form and any adsorption that may occur. Leaching of Sr is typically one to two orders of magnitude slower than that of Cs because of reactions between Sr and components of the cement. Minor components of the waste can significantly alter the leaching behaviour of radionuclides. For example, Co from decontamination waste streams may be complexed so that it is anionic and highly mobile even in the chemical environment of cement.

At present several countries are attempting to model the most important release mechanisms of radionuclides to establish an improved basis for extrapolations of leaching data to long times. Classical one dimensional diffusion experiments and leaching experiments are being used to validate model equations [106]. An accelerated leach test has been developed with an associated computer program that is used to determine if diffusion, solubility or diffusion plus adsorption is the controlling release mechanism [103, 105, 107]. Comparisons have been made between measured element concentration profiles in deteriorating cylindrical cement–NaNO$_3$ samples and concentration profiles calculated on the basis of finite element theory [108]. The leaching properties of cement based waste forms were examined by a combination of leaching and diffusion experiments and solid phase analysis to determine the intrinsic properties of materials that control leaching [109]. Calculations are being made to assess the effects of slowly flowing water in contact with waste forms on the near field chemistry and the resulting thermodynamic equilibrium concentrations of radionuclides [110, 111].

Speciation of transuranic elements is being investigated in several countries to determine their valence states and ionic or colloidal form. The colloid effects are likely to be important although complicated, mainly because of the complex structure and the composition and chemistry of cement systems. In particular, the solubility of the various structural components of the cement may be considerable in the pore water and then decline with a decrease in pH. Thus, soluble silicate may polymerize to yield colloids incorporating radionuclides. Other cement additives can also contribute to pseudocolloid formation if they are prone to dissolution in the alkaline pore water [112].

7.2.3. Waste form–container interactions

The integrity of the container influences the availability of a waste form for leaching and its degradation on exposure to water. Reactions of the waste form with the container material are a potential cause of premature container failure. To minimize this possibility, as well as to provide safety during storage, transportation and disposal, the quantity of free water in the container should be as small as possible. Moreover, the pH of the water should be basic to reduce the corrosion rate of steel or concrete containers. Testing for free water may consist of boring a small hole in the bottom of the container and measuring the drainable liquid. The simplicity of this method is offset by the obvious possibility of contamination. Other tests, such as those using acoustic sensing, should be considered. However, these tests should not be performed unless there is assurance that the waste forms have reached their prescribed cure time.

7.2.4. Waste form–disposal environment interactions

Chemical reactions between components of the waste form and the disposal environment are an important consideration for the durability of the waste form. Studies to determine the processes involved, and their rates, are site specific and often require a broad interdisciplinary approach. The reactive nature of cement based materials requires an understanding of the mechanisms involved in reactions between waste forms and the disposal environment. Ultimately, mathematical models can be developed to predict the long term behaviour of waste forms in the disposal environment.

One approach to studying long term reactions between waste forms and the disposal environment is to use lysimeters containing a waste form in an enclosed column of soil. The lysimeter is exposed to natural rainfall and other prevailing site conditions. A system of pumps and samplers is used to recover water percolating through the soil or collecting at the bottom. This leachate is analysed for radionuclides and other elements being released from the waste form over time, providing a realistic experiment with which near field leachate reactions can be studied. Lysimeter studies

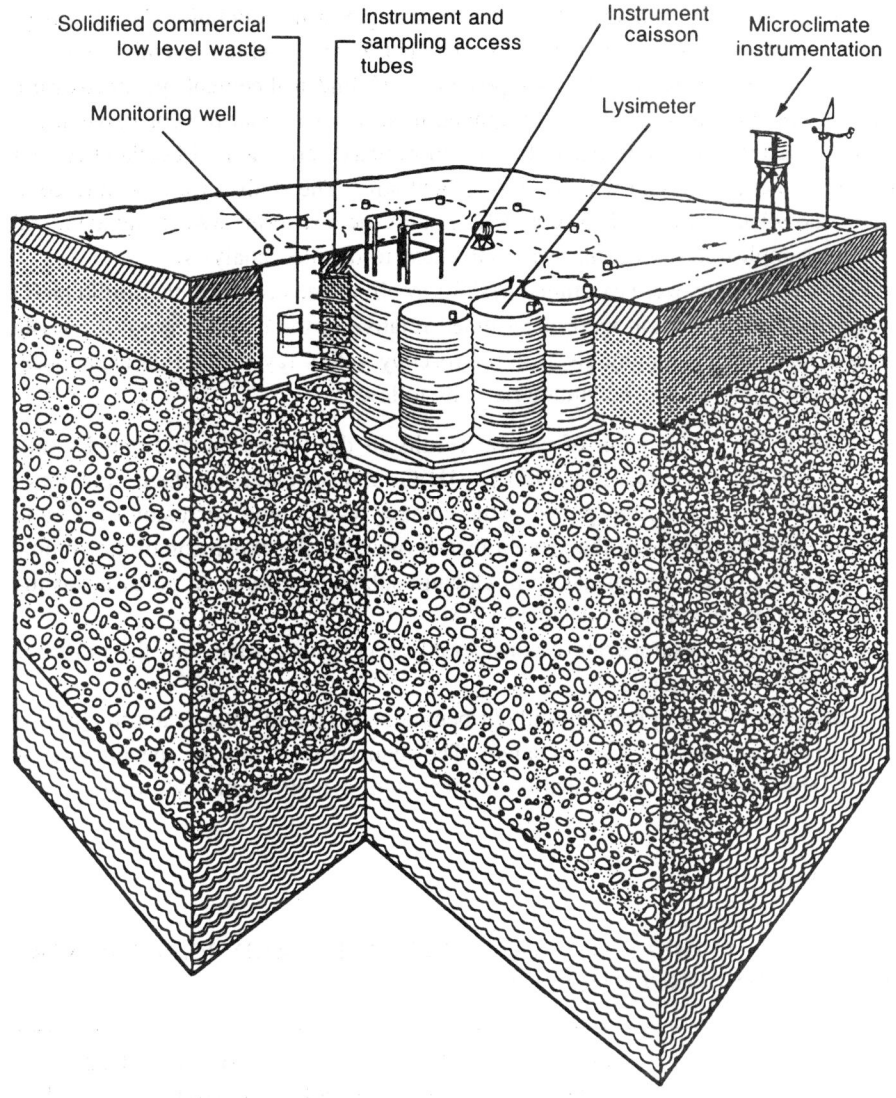

Solidified commercial low level waste

Instrument and sampling access tubes

Instrument caisson

Microclimate instrumentation

Monitoring well

Lysimeter

FIG. 32. Conceptual drawing of a lysimeter facility [116].

are presently being conducted in several locations, including the USA [113, 114] and Canada [115]. A conceptual drawing of a lysimeter installation is shown in Fig. 32 [116].

In Germany, studies have been conducted on the long term assessment of waste forms where consideration must be given to matrix dissolution and corrosion processes caused by reactions between the cement matrix and brine.

A series of experiments was conducted using a brine containing 24 wt% magnesium chloride and 2 wt% magnesium sulphate (Q brine), which resembles the corrosive brines occurring in rock-salt repositories. Cylindrical cement specimens were used, to which a plastic coating was applied in such a way that the corrosive attack was possible from the front face only. The specimens were made of Portland cement and slag cement, with and without 5 wt% bentonite. The studies were performed at 40, 55 and 90°C. After 1, 3, 5, 6 and 12 months the test series was terminated, and both the samples and the brine were subjected to detailed analysis.

The specimens were examined first for swelling and cracking, then solid phase and chemical composition were determined relative to the distance from the waste form surface using scanning electron microscopy and X ray diffraction techniques. The results obtained can be summarized as follows:

(a) The Q brine corrosion of cement results in the formation of Friedel's salt, $C_3A \cdot CaCl_2 \cdot 10H_2O$, $Mg_2(OH)_3Cl \cdot 4H_2O$ and calcium sulphate dihydrate, which partly seals the pores of the cement. Continuing corrosion eventually causes the collapse of the cement matrix owing to the expansion of the volume associated with the formation of calcium sulphate hydrate. In no case was the formation of ettringite detected [108].

(b) An increase in temperature considerably accelerates corrosion; the influence of pressure is small.

(c) Specimens of slag cement without bentonite are the most resistant to chloride penetration and Ca–Mg exchange. The reduced resistance of specimens containing bentonite can probably be attributed to their higher w/c ratios.

TABLE XXI. COMPARISON OF CALCULATED AND EXPERIMENTAL EQUILIBRIUM CONCENTRATIONS

	Cs-137 (nmol/L)	Sr-90 (nmol/L)	Np-237 (nmol/L)	Am-241 (nmol/L)	Pu-239/240 (pmol/L)
Experimental values	—	15	210	5	130
Calculated values[a]	37	31	40	8	960
Ratio of calculated to experimental values	—	2.1	0.2	1.6	7.4

[a] Based on an average level of Cs-137 of 6×10^{-8}M in the leachate and the following values for the distributions (K_d) of the nuclides between water and cement [107]: Cs: 5; Sr: 10; Np: 1000; Am: 5000; Pu: 8000.

In disposal sites where the rates of water flow are low, a thermodynamic approach may be used to describe the radionuclide releases and reactions between waste forms and the environment. The equilibrium concentrations resulting from long term waste form–leachate contact represent the maximum concentrations that nuclides can reach. The equilibrium concentrations are given by solubilities or sorption effects.

Measurements of equilibrium concentrations in the presence of repository constituents are being conducted at Harwell, UK. The research involves the determination of equilibrium values for U, Pu, Np and Am [110]. The solubilities of these elements in the presence of different anions (SO_4^{2-}, CO_3^{2-}) and their degrees of sorption on cement were determined. Also, the theoretical equilibrium concentrations of Cs, Sr, Np, Am and Pu were calculated. Calculated and experimental results for each isotope agree to within an order of magnitude, as shown in Table XXI. This indicates that:

— The average values obtained with Cs, Sr, Np, Am and Pu can be regarded as equilibrium values.
— Sorption is important in determining equilibrium values, since this maintains the levels of Pu, Am and Np in solution below their solubility limits.

The objectives of equilibrium chemistry studies are to obtain experimental data which will show how radionuclides will behave in the repository environment and to understand the processes which will control the aqueous concentration of the important radionuclides. This knowledge will allow the construction of models that can predict the behaviour of the repository over extended periods.

7.3. MECHANICAL STABILITY

One of the most relevant mechanical properties applicable to cement based material is its compressive strength. High compressive strength corresponds to superior long term chemical stability of the waste form. The compressive strength of cement waste forms depends on a set of parameters, of which the most important are:

— Water to cement ratio,
— Cement to waste ratio,
— Type of waste,
— Cure time.

The effects of these parameters have been discussed earlier in this report.

Standard compressive strength tests have been developed for cement mortars and concretes. Care should be taken to ascertain that the testing method used is appropriate for the mode of failure (e.g. plastic or brittle) of the waste form tested.

TABLE XXII. COMPRESSIVE STRENGTH (kPa) OF ORDINARY PORTLAND CEMENT AND POZZOLANIC CEMENT CONTAINING SIMULATED SULPHATE WASTE [117, 118]

$\dfrac{\text{water} + \text{salt}}{\text{cement}}$	Cured for 28 days (r.h. 95%)	Immersed in water for 1 month	Heated at 110°C for 7 days
Ordinary Portland cement			
0.37	556	590	850
0.41	528	562	768
0.44	488	520	710
0.48	422	470	604
Pozzolanic cement			
0.37	441	468[a]	—
0.41	413	446[a]	—
0.44	396	418[a]	—
0.48	364	392[a]	—

[a] Immersed in sea water.

The numerical specifications indicating acceptable compressive strength will vary depending on the specific disposal option. For example, if overburden pressure in near surface disposal is the only concern, perhaps 345 kPa (50 lb/in^2) is an acceptable value. However, if compressive strength is considered as an indicator of the durability of a waste form, significantly higher values are appropriate. Compressive strength is often used to evaluate the degradation that a sample undergoes after exposure to an environmental condition such as water immersion or freeze–thaw cycling. Results of compressive strength tests on replicate samples typically have a scatter of some 10% about a mean value. Consequently, a minimum of three to five samples is necessary for testing.

Tables XXII and XXIII illustrate the dependence of compressive strength on waste loading for sulphate and borate concentrates under various test conditions.

TABLE XXIII. COMPRESSIVE STRENGTH (kPa) OF ORDINARY PORT-
LAND CEMENT AND POZZOLANIC CEMENT CONTAINING BORATE
WASTE [117, 118]

$\dfrac{\text{water} + \text{salt}}{\text{cement}}$	Cured for 28 days (r.h. 90%)	Immersed in water for 1 month	Heated at 110°C for 7 days
Ordinary Portland cement			
0.46	246	270	145
0.52	212	236	132
0.58	187	204	126
0.63	179	196	125
Pozzolanic cement			
0.46	292	318[a]	—
1.52	258	282[a]	—
0.58	238	254[a]	—
0.63	230	246[a]	—

[a] Immersed in sea water.

7.4. THERMAL STABILITY

7.4.1. Degradation on heating

Cement as an inorganic binding material is not flammable. However, cement
waste forms can be degraded by removal of water at elevated temperatures. In addition, organic materials contained in a cement waste form (e.g. ion exchange resins
or combustible trash) can be decomposed by pyrolysis. The release of water or pyrolysis products from cement waste forms can cause the release of radioactivity, for
example by the formation of radioactive aerosols.

In principle, hydrated cement contains two different forms of water. The free
or unbound water is easily removed from the cement by heating to a temperature of
105°C. The other form is bound in the various hydrated phases and as interlayer
water and is released at higher temperatures depending on the decomposition temperatures of the hydrated phases.

TABLE XXIV. EFFECTS OF FREEZE–THAW CYCLES ON CEMENT PASTE
PLUS SIMULATED SULPHATE WASTE (ASTM C666)

water + salt / cement	w/c	Test I, 10 cycles	Test II, 10 cycles	Test III, cracks after:
Ordinary Portland cement				
0.30	0.30	Durable	Durable	50 cycles
0.37	0.30	Durable	Durable	40 cycles
0.41	0.33	Durable	Durable	35 cycles
0.44	0.36	Durable	Durable	35 cycles
0.48	0.39	Durable	Durable	35 cycles
Pozzolanic cement				
0.30	0.30	Durable	Durable	10 cycles
0.37	0.30	Durable	Cracked	—
0.41	0.33	Durable	Cracked	—
0.44	0.36	Durable	Cracked	—
0.48	0.39	Durable	Cracked	—

Thermogravimetric analysis (TGA) is helpful for interpreting the phenomena which take place during heating. With cement products, the following observations were made [119]:

25–200°C: Loss of sorbed, capillary and crystallization water;

25–800°C: Continuous release of water from the calcium silicate hydrate (CSH) and calcium aluminate hydrate (CAH) phases;

350–450°C: Decomposition of calcium hydroxide;

500–800°C: Decomposition of calcium carbonate.

7.4.2. Freeze–thaw stability

For long term storage on sites where low temperatures can be experienced, the resistance of waste forms to freeze–thaw cycling should be assessed.

The freezing behaviour of water in gel pores and capillary water is basically different, the susceptibility of the cement to freezing being a function of w/c, waste loading and air porosity. Freezing can cause an unpredictable expansion as a result of a wide distribution in capillary pores [120].

The most widely used test of the resistance of concrete to freezing is the ASTM test "Resistance of Concrete to Rapid Freezing and Thawing" (ASTM C666). The cyclic frequency and the temperature gradients applied should be carefully selected for testing. The permissible freezing rate of 6–60°C/h in these tests differs from natural conditions, where the rate seldom exceeds 3°C/h [121].

Table XXIV gives data on freeze–thaw cycling from tests on waste forms containing simulated sulphate waste.

7.5. RADIATION STABILITY

The absorption of radiation from the incorporated radionuclides in the waste form may change the properties of the waste form. Cements and concrete are generally regarded as resistant to radiation doses of the order of 10^8 Gy. At this dose, no changes in mechanical properties were observed [122].

Two types of damage from radiation can occur in encapsulated ILW: atomic displacements resulting from particle interactions and chemical effects resulting from radiolysis of loosely bound molecules. These atomic displacements are unlikely to be of importance over the time-scale of interest because of the low concentration of α emitters in ILW.

Most of the work has been carried out on simulated waste using an external γ radiation source. The effects of γ radiation at high dose rates on dimensional stability, mechanical properties, gas evolution and leaching have been studied. However, little information exists on the effects of α and β radiation on waste forms.

A major effect of radiation is radiolysis gas formation. During γ irradiation, most waste forms evolve hydrogen, and oxygen is absorbed. The quantity of hydrogen evolved appears to be a linear function of absorbed dose up to the maximum of 9 MGy [123]. $G(H_2)$ values (molecules/100 eV) for waste forms tend to lie between 0.05 and 0.13, although values as low as 0.01 have been observed [124].

7.6. MICROBIAL DEGRADATION

Microbiota cannot exist within concrete because of the high alkalinity and the lack of nutrients [125]. Bacteria and fungi can degrade concrete structures under certain conditions but it is uncommon to find any more than superficial effects. Bacteria can indirectly degrade concrete through chemical reactions of concrete with products of bacterial metabolism. This is observed in sewer systems where the concrete pipes are degraded. Anaerobic bacteria reduce sulphate and organic sulphur compounds in the sewage to form H_2S gas and volatile organic sulphur compounds. These are released and migrate to the top of the sewer pipes where they are biologically oxidized to form sulphuric acid, which attacks the concrete pipe [125]. This type of degradation can occur rapidly, causing failure in a period of the order of ten years after installation.

Fungi can also, in rare cases, cause degradation of concrete through the acidity (pH2) of fluids contained in their mycelia [126]. Laboratory tests indicate that some fungi can penetrate 3 cm of cement mortar in 3–4 months.

Cement based low and intermediate level radioactive waste forms are unlikely to be attacked by microbiota to any significant extent. While the waste contained in the cement may serve as a source of nutrients, the large excess of nutrients found in sewage is not available in disposal sites. In addition, the chemical and physical conditions necessary for the formation of large quantities of sulphuric acid (a closed system with anaerobic conditions at one end and aerobic conditions at the other) are generally not present. When selecting tests to discern microbial growth, care should be taken to find tests that are intended to observe actual degradation processes. Tests meant to determine if properties such as optical clarity or colour are altered by microbiota are probably not good indicators of the durability of cement under microbial attack.

Microorganisms change the groundwater chemistry at disposal sites by altering the pH and Eh [127]. These changes affect speciation of some radionuclides, influencing their mobility and the stability of compounds in the disposal environment. While this level of microbial activity may affect chemical reactions at the surface of waste forms, it would not be expected to affect waste form durability.

REFERENCES

[1] INTERNATIONAL ATOMIC ENERGY AGENCY, Standardization of Radioactive Waste Categories, Technical Reports Series No. 101, IAEA, Vienna (1970).

[2] INTERNATIONAL ATOMIC ENERGY AGENCY, Conditioning of Low- and Intermediate-Level Radioactive Wastes, Technical Reports Series No. 222, IAEA, Vienna (1983).

[3] INTERNATIONAL ATOMIC ENERGY AGENCY, Treatment of Low- and Intermediate-Level Liquid Radioactive Wastes, Technical Reports Series No. 236, IAEA, Vienna (1984).

[4] INTERNATIONAL ATOMIC ENERGY AGENCY, Treatment of Spent Ion-Exchange Resins for Storage and Disposal, Technical Reports Series No. 254, IAEA, Vienna (1985).

[5] INTERNATIONAL ATOMIC ENERGY AGENCY, Immobilization of Low and Intermediate Level Radioactive Wastes with Polymers, Technical Reports Series No. 289, IAEA, Vienna (1988).

[6] PHILLIPS, J., FEIZOLLAHI, F., MARTINEIT, R., BELL, W., STOVKY, R., A Waste Inventory Report for Reactor and Fuel-Fabrication Facility Wastes, Rep. ONWI-20, NUS Corp., Rockville, MD (1979).

[7] ELECTRIC POWER RESEARCH INSTITUTE, Low-Level Radwaste Solidification, Rep. EPRI-NP-2900, Palo Alto, CA (1983).

[8] BENNET, D., BRADBURY, D., KAYE, C.S., MEESE, R., Conditioning Optional for Magnox Fuel Element Debris, Radioactive Waste Management (Proc. Conf. London, 1984), British Nuclear Energy Society, London (1985) 114.

[9] POTTIER, P.E., GLASSER, F.P. (Eds), Characterization of Low- and Medium-Level Radioactive Waste Forms, EUR-10579-EN, Commission of the European Communities, Luxembourg (1986).

[10] SOROKA, I., Portland Cement Paste and Concrete, 1st edn, Chemicals Publishing, New York (1980).

[11] LEA, F.M., The Chemistry of Cement and Concretes, 3rd edn, Edward Arnold, Glasgow (1970).

[12] MINDES, S., YOUNG, J.F., Concrete, Prentice-Hall, Englewood Cliffs, NJ (1981).

[13] AMERICAN SOCIETY FOR TESTING AND MATERIALS, Compound Composition Limits for Cement, Spec. C150-78, ASTM, Philadelphia, PA (1978).

[14] NEVILLE, A.M., Properties of Concrete, 3rd edn, Pitman, London (1981).

[15] FORRESTER, J.A., A conduction calorimeter for the study of cement hydration, Cem. Technol. (May–June 1970).

[16] WOODS, H., Observations on the resistance of concrete to freezing and thawing, J. Am. Concr. Inst. 51 (1954) 345–349.

[17] NURSE, R.W., "Slag cements", The Chemistry of Cements (TAYLOR, H.F.W., Ed.), Vol. 2, Academic Press, New York (1984) 37–68.

[18] PALMER, J.D., SMITH, D.L.G., The Incorporation of Low- and Medium-Level Radioactive Wastes (Solids and Liquids) in Cement, EUR-10561-EN, Commission of the European Communities, Luxembourg (1986).

[19] BROWN, O., LEE, D.J., PRICE, M.S.T., SMITH, D.L.G., "Cement based processes for the immobilization of intermediate-level waste", Radioactive Waste Management, British Nuclear Energy Society, London (1985).

[20] GLASSER, F.P., McCULLOCH, C., Characterization of Radioactive Waste Forms, Progress Report for 1986, EUR-11354, Commission of the European Communities, Luxembourg (1986).

[21] LAHUMA, H., Quelques aspects de la physico-chimie des ciments alumineux, Rev. Gén. Sci. Appl. 1 3 (1952) 66–74.

[22] NEVILLE, A.M., Tests on the strength of high-alumina cement concrete, J. N. Z. Inst. Eng. 14 3 (1959) 73–76.

[23] MIDGLEY, H.G., The mineralogy of set high-alumina cement, Trans. Br. Ceram. Soc. 66 4 (1967) 161–187.

[24] HUSSEY, A.V., ROBSON, T.D., "High alumina cement as a structural material in the chemical industry", Proc. Symp. on Materials of Construction in the Chemical Industry, Birmingham, Society of Chemical Industry (1950).

[25] CLARK, D.E., COLOMBO, P., NEILSON, R.M., Jr., Solidification of Oils and Organic Liquids, Rep. BNL-51612, Brookhaven Natl Lab., Upton, NY (1982).

[26] BROWNSTEIN, B., LEVESQUE, R.G., Experience with Cement Usage as the Binding Agent for Radwaste, Tech. Paper 78-NE-15, American Society of Mechanical Engineers, New York (1978).

[27] ZHOU, H., COLOMBO, P., Solidification of Low-Level Radioactive Waste in Masonry Cement, Rep. BNL-52074, Brookhaven Natl Lab., Upton, NY (1987).

[28] ROSENSTIEL, T.L., LANGE, R.G., "The solidification of low-level radioactive organic fluids with Envirostone gypsum cement", Waste Management '84 (Proc. Symp. Tucson, 1984) (POST, R.G., WACKS, M.E., Eds), Vol. 2, Low-Level Waste, Volume Reduction Methodologies and Economics, Arizona Board of Regents, Tucson (1984).

[29] POLIVKA, M., WILSON, C., Proc. Symp. on Expansive Cement Concrete, Rep. SP-38, American Concrete Inst., Detroit, MI (1973) 235.

[30] AMERICAN CONCRETE INSTITUTE COMMITTEE 223, Expansive cement concrete — Present state of knowledge, J. Am. Concr. Inst. 67 8 (1970) 583–610.

[31] AMERICAN CONCRETE INSTITUTE COMMITTEE 223, Recommended practice for the use of shrinkage-compensating cement, J. Am. Concr. Inst. 73 6 (1976) 319–339.

[32] UCHIKAWA, H., ISUKIYAMA, K., The hydration of jet cement at 20°C, Cem. Concr. Res. 3 3 (1973) 263–277.

[33] PERENCHIO, W., "Regulated-set cements — Application and field problems", New Materials in Concrete Construction (Proc. Conf. 1971) (SHAH, S.P., Ed.), Univ. of Illinois, Chicago (1972) 12.

[34] RUDOLPH, L., LUO, S., VEJMELKA, P., KÖSTER, R., Untersuchungen zum Abbindeverhalten von Zementsuspensionen, Rep. KFK-3401, Kernforschungs-zentrum Karlsruhe (1982).

[35] RUSSELL, P., Concrete Admixtures, Eyre & Spottiswoode, Andover, UK (1983).

[36] Entrained Air in Concrete: A Symposium, ACI J., Proc. 42 6 (1946) 610–700.

[37] BLANKS, R.F., CORDON, W.A., Practices, experiences, and tests with air-entraining agents in making durable concrete, ACI J., Proc. **45** 6 (1949) 469–488.

[38] PORTLAND CEMENT ASSOCIATION, Design and Control of Concrete Mixtures, 12th edn, PCA, Skokee, IL (1979).

[39] BUREAU OF RECLAMATION, Concrete Manual, 8th edn, Bureau of Reclamation, Denver, CO (1975).

[40] AMERICAN CONCRETE INSTITUTE COMMITTEE 211, Recommended Practice for Selecting Proportions for Normal and Heavyweight Concrete, ACI, Detroit, MI (1977).

[41] POWERS, T.C., Topics in concrete technology, Part 3. Mixtures containing intentionally entrained air, J. PCA Res. Dev. Lab. **6** 3 (1964) 19–42.

[42] BRUERE, S.M., J. Appl. Chem. Biotechnol. **21** 3 (1971) 61–64.

[43] POWERS, T.C., Properties of Fresh Concrete, Wiley, New York (1968).

[44] JACKSON, F.H., TIMMS, A.G., Evaluation of air-entraining admixtures for concrete, Public Roads (1954) 259–267.

[45] KANTRO, D.L., Tricalcium silicate hydration in the presence of various salts, J. Test. Eval. **3** 4 (1975) 312–321.

[46] LIEBER, W., RICHARTZ, W., Effects of triethanolamine, sugar and boric acid on the setting and hardening of cements, Zem.–Kalk–Gips **25** 9 (1972) 403–409 (in German).

[47] RAMACHANDRAN, V.S., Action of triethanolamine on the hydration of tricalcium aluminate, Cem. Concr. Res. **3** 1 (1973) 41–54.

[48] EDWARDS, L.C., AUGSTADT, R.L., The effect of some soluble inorganic admixtures on the early hydration of Portland cement, J. Appl. Chem. Biotechnol. **16** 5 (1966) 166–168.

[49] MURAKAM, J., TANAKA, G., "Contribution of calcium thiosulfate to the acceleration of the hydration of Portland cement and comparison with other soluble inorganic salts", Chemistry of Cement (Proc. 5th Int. Symp. Tokyo, 1968), Vol. 2, Cement Association of Japan, Tokyo (1969) 422–436.

[50] Admixtures for Mortar and Concrete (Proc. RILEM-ABEM Int. Symp. Brussels, 1967).

[51] ROSSKOPF, P.A., LINTON, F.J., PEPPLER, R.B., Effect of various accelerating chemical admixtures on setting and strength development of concrete, J. Test. Eval. **3** 4 (1975) 322–330.

[52] STEIN, H.M., Influence of some additives on the hydration reactions of Portland cement, II. Electrolytes, J. Appl. Chem. Biotechnol. **2** 12 (1961) 482–492.

[53] WASHA, S.W., WITHEY, N.H., Strength and durability of concrete containing Chicago fly ash, ACI J., Proc. **49** 8 (1953) 701–712.

[54] ARBER, M.G., VIVIAN, H.E., Inhibition of the corrosion of steel imbedded in mortars, Aust. J. Appl. Sci. **12** 12 (1969) 339–347.

[55] BASH, S.M., RAKIMBAEV, Sh.M., Quick-setting cement mortars containing organic additives, Beton Zhelezobeton **15** 7 (1969) 44–45 (in Russian).

[56] KOSSIVAS, T.G., Setting Accelerators for Portland Cement, German Patent 2 114 081, 14 Oct. 1971.

[57] BURERE, L.M., NEBEGIN, J.D., WILSON, L.M., A Laboratory Investigation of the Drying Shrinkage of Concrete Containing Various Types of Chemical Admixtures, Tech. Paper No. 1, Div. of Applied Mineralogy, Commonwealth Scientific and Industrial Res. Org., East Melbourne, Australia (1971).

[58] FERET, L., VERMAT, N., The effect on shrinkage and levelling of mixing different cements to obtain rapid set, Rev. Matér. Constr. No. 496 (1957) 1–10.

[59] BALAZS, B., KELMEN, J., KILIAN, J., New method for accelerating the hardening of concrete, Epitoeanyag **10** 9 (1958) 326–331.

[60] DURIEX, M., LEZY, R., New possibilities for ensuring the rapid hardening of cements, mortars and concretes, Ann. Inst. Tech. Bâtim. Trav. Publics **9** (1956) 138.

[61] LADO, E., Method of Producing Concrete of Improved Strength, US Patent 3 389 003, 18 June 1968.

[62] AUGSTADT, R.L., HURLEY, F.R., Spodumene as an Accelerator for Hardening Portland Cement, US Patent 3 331 695, 18 July 1967.

[63] KROONE, B., Reaction between hydrating Portland cement and ultramarine blue, Chem. Ind. (London) (2 Mar. 1968) 287–288.

[64] VAGNOV, C.R., GRANKOVSKII, S.G., KRUGLITZKI, N.N., OVCHINNIKOVA, A.T., NADD, L.G., Interaction of cement with palgorskite during hydration, Khim. Tekhnol. Vody (1 Nov. 1972) 13–18 (in Russian).

[65] NELSON, J.A., YOUNG, J.F., Addition of colloidal silicas and silicates to Portland cement pastes, Cem. Concr. Res. **7** 3 (1977) 277–282.

[66] STEIN, H.M., STEVELS, J.M., Influence of silica on the hydration of tricalcium silicate, J. Appl. Chem. **14** 8 (1974) 338–346.

[67] RAMACHANDRAN, V.S., "Calcium chloride in concrete", Science and Technology, Applied Science Publishers, London (1976).

[68] TENOUTASSE, N., "The hydration mechanism of C3A and C3S in the presence of calcium chloride and calcium sulfate", Chemistry of Cement (Proc. 5th Int. Symp. Tokyo, 1968), Vol. 2, Cement Association of Japan, Tokyo (1969) 372–378.

[69] PERENCHIO, W.F., WHITING, D.A., KANTRO, D.L., Superplasticizers in Concrete (Int. Proc. 1), Canada Centre for Mineral and Energy Technology, Ottawa (1978).

[70] RAMACHANDRAN, V.S., FELDMAN, R.F., BEAUDOIN, J.J., Concrete Science, Heyden, London (1981).

[71] MASSAZZA, F., "Chemistry of pozzolanic additions and mixed cements", Chemistry of Cement (Proc. 6th Int. Congr. Moscow, 1974).

[72] SERSALE, R., "Structure and characteristics of pozzolans and fly ashes", Chemistry of Cement (Proc. 7th Int. Congr. Paris, 1980).

[73] Chemistry of Cement (Proc. 7th Int. Congr. Paris, 1980), Vol. 1.

[74] BERRY, E.E., MALHOTRA, V.M., Fly ash for use in concrete — A critical review, ACI J. **2** 3 (1981) 40–52.

[75] HOGAN, F.J., MEUSEL, J.W., Evaluation for durability and strength development of a ground granulated blast furnace slag, Cem. Concr. Aggr. **3** 1 (1981) 40–52.

[76] FROHNSDORFF, G., CLIFTON, J.R., Fly Ash in Cement and Concrete — Technical Needs and Opportunities, Rep. NBSIR 81-2239, Natl Bureau of Standards, Washington, DC (1981).

[77] MEHTA, P.K., Studies of blended cements containing Santorin earth, Cem. Concr. Res. **11** 4 (1981) 507–518.

[78] MEHTA, P.K., "Pozzolanic and cementitious byproducts as mineral admixtures for concrete — A critical review", Use of Fly Ash, Slags and Silica Fume in Concrete (Proc. Int. Conf. Montebello), Rep. SP-79, ACI, Detroit, MI (1983).

[79] GUTT, W., NIXON, P.J., Use of waste materials in the construction industry — Analysis of the RILEM symposium by correspondence, Mater. Struct. **12** 70 (1979) 255–306.

[80] BERRY, E.E., MALHOTRA, V.M., Fly ash for use in concrete — A critical review, J. Am. Concr. Inst. **77** 8 (1980) 59–73.

[81] LANE, R.O., BEST, J.F., Properties and use of fly ash in Portland cement concrete, Concr. Int. **4** 7 (1982) 81–92.

[82] DAVIS, R.E., CARLSON, R.W., KELLY, J.W., DAVIS, H.E., Properties of cement and concrete containing fly ash, ACI J., Proc. **33** 5 (1937) 577–612.

[83] MEHTA, P.K., PITT, N., Energy and industrial materials from crop residues, Resour. Recovery Conserv. **2** (1976) 23–38.

[84] HEGGINSON, E.C., Significance of Tests and Properties of Concrete and Concrete-Making Materials, ASTM-STP-169A, American Society for Testing and Materials, Philadelphia, PA (1966) 543–555.

[85] WAGNER, H.B., Polymer-modified hydraulic cements, Ind. Eng. Chem. Prod. Res. Dev. **4** 3 (1965) 191–196.

[86] OHAMA, Y., Study on Properties and Mix Proportioning of Polymer-Modified Mortars for Buildings, Rep. 65, Building Research Inst., Tokyo (1973) (in Japanese).

[87] SCHWEITE, H.E., LUDWIG, V., AACHEN, G.S., The influence of plastics dispersions on the properties of cement mortars, Betonstein Ztg. **1** 35 (1969) 7–16.

[88] WAGNER, H.B., GRENLEY, D.G., Interphase effects in polymer-modified hydraulic cements, J. Appl. Polym. Sci. **22** 3 (1978) 821–822.

[89] FUHRMANN, M., NEILSON, R.M., Jr., COLOMBO, P., A Survey of Agents and Techniques Applicable to the Solidification of Low-Level Radioactive Wastes, Rep. BNL-51521, Brookhaven Natl Lab., Upton, NY (1981).

[90] NEILSON, R.M., Jr., COLOMBO, P., Solidification of Ion-Exchange Resin Wastes, Rep. BNL-51615, Brookhaven Natl Lab., Upton, NY (1982).

[91] PATEK, P., Study of Mechanical and Physico-Chemical Properties of Cemented Spent Ion-Exchange Resins, Progress Report of a Co-ordinated Research Programme of the IAEA, 1981.

[92] NUCLEAR REGULATORY COMMISSION, Standard Review Plan for Review of Safety Analysis Reports for Nuclear Power Plants, LWR edn, NRC, Washington, DC (1981).

[93] JOLLEY, R.L., et al., Low-Level Radioactive Waste from Commercial Nuclear Reactors, Rep. ORNL/TM-9846, Vol. 2, Oak Ridge Natl Lab., TN (1986).

[94] CHRISTENSEN, H., "Cement solidification of BWR and PWR radioactive waste at the Ringhals Nuclear Power Plant", On-Site Management of Power Reactor Wastes (Proc. OECD/NEA–IAEA Symp. Zurich, 1979), Nuclear Energy Agency of the OECD, Paris (1979).

[95]   AMERICAN SOCIETY OF MECHANICAL ENGINEERS, Radioactive Waste Technology, ASME, New York (1980).

[96]   ELECTRIC POWER RESEARCH INSTITUTE, Low-Level Radwaste Solidification, Rep. EPRI-NP-2900, Palo Alto, CA (1983) pp. 6-6, Appx E-5, Appx E-6.

[97]   GUYMON, R.H., VANSELOW, L.D., CAMPBELL, G.D., "The Grout Treatment Facility. Processing facilities for low-level waste immobilization and disposal", Spectrum '88 (Proc. Int. Topical Mtg on Nuclear and Hazardous Waste Management), CONF-880903, American Nuclear Society, La Grange Park, IL (1988) 95-98.

[98]   WILHITE, E.L., "Concept development for saltstone and low-level waste disposal", Waste Management '87 (Proc. Conf. Tucson, 1987) (POST, R.G., Ed.), Vol. 2, Univ. of Arizona, Tucson (1987) 63-68.

[99]   NUCLEAR REGULATORY COMMISSION, Final Waste Classification and Waste Form Technical Position Papers, Rep. NRC-4900815, Washington, DC (1983).

[100]   MATERIALS CHARACTERIZATION CENTER, MCC-1, Static Leach Test, Pacific Northwest Lab., Richland, WA (1980).

[101]   HESPE, E.D., Leach testing of immobilized radioactive waste solids, A proposal for a standard method, At. Energy Rev. 9 1 (1971) 195-207.

[102]   INTERNATIONAL ORGANIZATION FOR STANDARDIZATION, Draft ISO Standard, Long-Term Leach Testing of Radioactive Waste Solidification Products, ISO/TC85/SC5/WG.

[103]   FUHRMANN, M., HEISER, J.H., PIETRZAK, R.F., COLOMBO, P., Accelerated Leach Test (draft), Brookhaven Natl Lab., Upton, NY (1989).

[104]   MATERIALS CHARACTERIZATION CENTER, MCC-4s, Low-Flow-Rate Leach Test Method, Pacific Northwest Lab., Richland, WA (1981).

[105]   FUHRMANN, M., PIETRZAK, R., HEISER, J., FRANZ, E.M., COLOMBO, P., "The effects of temperature on the leaching behavior of cement waste forms: The cement/sodium sulfate system", Scientific Basis for Nuclear Waste Management (Proc. 13th Symp. Boston, 1989), CONF-891119-20, Rep. BNL-43449 (1989).

[106]   DOZOL, M., KRISCHER, W., POTTIER, P., SIMON, R. (Eds), Leaching of Low and Medium Level Waste Packages under Disposal Conditions, Rep. EUR-10220, Commission of the European Communities, Luxembourg (1985) 31.

[107]   FUHRMANN, M., PIETRZAK, R.F., FRANZ, E.M., HEISER, J.H., COLOMBO, P., Optimization of the Factors that Accelerate Leaching, Rep. BNL-52204, Brookhaven Natl Lab., Upton, NY (1989).

[108]   KIENZLER, B., KÖSTER, R., Experimental and theoretical investigations of corrosion mechanisms in cemented waste forms, Nucl. Technol. 71 (1985).

[109]   VAN DER SLOT, H., DE GROOT, G., WIZKSTRA, J., "Leaching characteristics of construction materials and stabilization products containing waste materials", Environmental Aspects of Stabilization and Solidification of Hazardous and Radioactive Wastes, STP 1033 (COTÉ, P.L., GILLIAM, T.M., Eds), American Society for Testing and Materials, Philadelphia, PA (1989) 125-149.

[110]   HODGKINSON, D.P., et al., The NIREX Safety Assessment Research Programme Annual Report for 1985/86, Rep. AERE-R-12331, Atomic Energy Research Establishment, Harwell (1987).

[111] CRISCENTI, L.J., SERNE, R.J., Geochemical Analysis of Leachate from Cement/Low-Level Radioactive Waste/Soil Systems, Rep. PNL-6544, Pacific Northwest Lab., Richland, WA (1988).

[112] RAMSAY, J.D.F., AVERY, R.G., Colloids Related to Low-Level and Intermediate-Level Waste, Rep. AERE-R-12538, Atomic Energy Research Establishment, Harwell (1987).

[113] OBLATH, S., HOEFFNER, S.L., "Evaluation and performance of the special waste form lysimeters at a humid site", Proc. 7th Annu. Participants' Information Mtg, Natl Low-Level Radioactive Waste Management Program, CONF-8509121, EG&G Idaho, Idaho Falls (1986).

[114] McCONNELL, J.W., Jr., ROGERS, R.D., Field testing of waste forms using lysimeters", Waste Management '87 (Proc. Conf. Tucson, 1987) (POST, R.G., Ed.), Univ. of Arizona, Tucson (1987) 551–556.

[115] BUCKLEY, L.P., TOSELLO, N.B., WOODS, B.L., "Leaching low-level radioactive waste in simulated disposal conditions", Environmental Aspects of Stabilization and Solidification of Hazardous and Radioactive Wastes (Proc. 4th Int. Symp. Atlanta, 1987) (COTÉ, P.L., GILLIAM, T.M., Eds), CONF-870511, American Society for Testing and Materials, Philadelphia, PA (1989) 330–342.

[116] WALTER, M.B., GRAHAM, M.J., "Special waste form lysimeters — arid", Proc. 7th Annu. Participants' Information Mtg, Natl Low-Level Radioactive Waste Management Program, CONF-8509121, EG&G Idaho, Idaho Falls (1986).

[117] ZHOU, H., COLOMBO, P., "Solidification of radioactive waste in a cement–lime mixture", Waste Management '84 (Proc. Symp. Tucson, 1984) (POST, R.G., WACKS, M.E., Eds), Arizona Board of Regents, Tucson (1984).

[118] LOKKEN, R.O., A Review of Radioactive Waste Immobilization in Concrete, Rep. PNL-2654, UC-70, Battelle Pacific Northwest Lab., Richland, WA (1978).

[119] MANNS, W., Über den Wassergehalt von Beton bei höheren Temperaturen, Betontech. Ber. 1/75 (1975).

[120] KREUKLER, K., Chemie des Bauwesens, Vol. 1, Anorganische Chemie, Springer-Verlag, Berlin (1980).

[121] RAMACHANDRAN, V.S., FELDMAN, R.F., BEAUDOIN, J.J., Concrete Sciences, Heyden, London (1981) 364.

[122] HILSDORF, H.K., KROPP, I., KOCH, H.J., Der Einfluss radioaktiver Strahlung auf die mechanischen Eigenschaften von Beton, Vol. 261, Deutscher Ausschuss für Stahlbeton, Berlin (1966).

[123] MÖCKEL, H.J., KÖSTER, R.H., Gas formation during the gamma-radiolysis of cemented low- and intermediate-level waste products, Nucl. Technol. **59** (1982) 414.

[124] STONE, J.A., Evaluation of Concrete as Matrix for Solidification of Savannah River Plant Waste, Rep. DP-1448, Savannah River Lab., Aiken, SC (1977).

[125] EGLINTEN, M.S., Concrete and its Chemical Behaviour, Thomas Telford, London, (1987).

[126] WOODS, H., Durability of Concrete Construction, ACI Monogr. 4, American Concrete Inst., Detroit, MI (1968).

[127] WEST, J.M., McKINLEY, J.G., CHAPMAN, N.A., Microbes in deep geological systems and their possible influence on radioactive waste disposal, Radioact. Waste Manage. Nucl. Fuel Cycle **3** 1 (1982) 1–15.

GLOSSARY

absorption. The process by which a liquid is drawn into and tends to fill permeable pores in a porous solid body; also the increase in weight of a porous solid body resulting from the penetration of a liquid into its permeable pores.

acceleration. Increase in velocity or in rate of change, especially the quickening of the natural progress of a process, such as hardening, setting or strength development of concrete.

accelerator. A substance which, when added to concrete, mortar or grout, increases the rate of hydration of the hydraulic cement, shortens the time of setting, or increases the rate of hardening or strength development, or both.

additive. A term frequently (but improperly) used as a synonym for an addition or admixture.

admixture. A material, other than water, aggregates and hydraulic cement, used as an ingredient of concrete or mortar, and added to the concrete immediately before or during its mixing.

adsorption. Development at the surface of a liquid or solid of a higher concentration of a substance than exists in the bulk of the medium; especially the formation of one or more layers of molecules of gases, of dissolved substances, or of liquids at the surface of a solid, such as cement, cement paste or aggregate, or of air entraining agents at the air–water interfaces; also the process by which a substance is adsorbed.

aggregate. Granular material, such as sand, gravel, crushed stone and iron blast-furnace slag, used with a cementing medium to form a hydraulic cement, concrete or mortar.

air entraining agent. An addition for hydraulic cement or an admixture for concrete or mortar which causes entrained air to be incorporated in the concrete or mortar during mixing, usually to increase its workability and frost resistance. (See **entrained air**.)

air entrainment. The occlusion of air in the form of minute bubbles (generally smaller than 1 mm) during the mixing of concrete or mortar.

alumina. Aluminium oxide (Al_2O_3).

aluminate concrete. Concrete made with calcium aluminate cement; used primarily where high early strength, refractory or corrosion resistant concrete is required.

argillaceous. Composed primarily of clay or shale, clayey.

batch mixer. A machine that mixes batches of concrete or mortar, in contrast to a continuous mixer.

bauxite. A rock composed principally of hydrous aluminium oxides; the principal ore of aluminium and a raw material for the manufacture of calcium aluminate cement.

bentonite. A clay composed principally of minerals of the montmorillonite group, characterized by high adsorption and very large volume change with wetting or drying.

blast-furnace slag. The non-metallic product, consisting essentially of silicates and aluminosilicates of calcium and other bases, that is developed in a molten condition simultaneously with iron in a blast-furnace.

    (a) Air cooled blast-furnace slag is the material resulting from solidification of molten blast-furnace slag under atmospheric conditions; subsequent cooling may be accelerated by applying water to the solidified surface.

    (b) Expanded blast-furnace slag is the lightweight, cellular material obtained by controlled processing of molten blast-furnace slag with water, or water and other agents, such as steam or compressed air, or both.

    (c) Granulated blast-furnace slag is the glassy, granular material formed when molten blast-furnace slag is rapidly chilled, as by immersion in water.

bulk density. The weight of a material (including solid particles and any contained water) per unit volume, including voids.

calcareous. Containing calcium carbonate or, less generally, containing the element calcium.

calcite. A mineral having the composition calcium carbonate ($CaCO_3$) and a specific crystal structure; the principal constituent of limestone, chalk and marble; used as a major constituent in the manufacture of Portland cement.

calcium aluminate cement. The product obtained by pulverizing clinker consisting essentially of hydraulic calcium aluminates resulting from fusing or sintering a suitably proportioned mixture of aluminous and calcareous materials; called high alumina cement in the UK.

calcium chloride. A crystalline solid, $CaCl_2$; in various technical grades, used as a drying agent, an accelerator of concrete and a de-icing chemical, and for other purposes.

calcium silicate hydrate. Any of the various reaction products of calcium silicate and water, often produced by curing in an autoclave.

carbonation. Reaction between carbon dioxide and a hydroxide or oxide to form a carbonate, especially in cement paste, mortar or concrete; the reaction with calcium compounds to produce calcium carbonate.

catalyst. A substance that initiates a chemical reaction and enables it to proceed under milder conditions than otherwise required and which does not itself alter or enter into the reaction.

celite. A name used to identify the calcium aluminoferrite constituent of Portland cement.

cement, high early strength. Cement characterized by producing earlier strength in mortar or concrete than regular cement; referred to in the USA as Type III.

cement, masonry. A hydraulic cement for use in mortars for masonry construction, containing one or more of the following materials: Portland cement, Portland blast-furnace slag cement, Portland pozzolan cement, natural cement, slag cement or hydraulic lime, and in addition usually containing one or more materials such as hydrated lime, limestone, chalk, calcareous shell, talc, slag or clay, prepared for this purpose.

cement, natural. A hydraulic cement produced by calcining a naturally occurring argillaceous limestone at a temperature below the sintering point and then grinding it to a fine powder.

cement, Portland. A hydraulic cement produced by pulverizing clinker consisting essentially of hydraulic calcium silicates, and usually containing one or more of the forms of calcium sulphate as an interground addition.

cement, Portland blast-furnace slag. A hydraulic cement consisting of an intimately interground mixture of Portland cement clinker and granulated blast-furnace slag or an intimate and uniform blend of Portland cement and finely granulated blast-furnace slag in which the amount of slag is within specified limits.

cement, slag. A hydraulic cement consisting essentially of an intimate and uniform blend of granulated blast-furnace slag and hydrated lime in which the slag constituent is more than a specified minimum percentage.

cement, sulphate resistant. Portland cement that is low in tricalcium aluminate, to reduce the susceptibility of concrete to attack by dissolved sulphates in water or soils; designated Type V in the USA.

cement, supersulphated. A hydraulic cement made by intimately intergrinding a mixture of granulated blast-furnace slag, calcium sulphate, and a small amount of lime, cement or cement clinker; so named because the equivalent content of sulphate exceeds that for Portland blast-furnace slag cement.

cement gel. The colloidal material that makes up the major portion of the porous mass of mature hydrated cement paste.

cementation process. The process of injecting cement grout under pressure into certain types of ground (e.g. gravel, fractured rock) to solidify it.

cementitious. Having cementing properties.

clay. Natural mineral material having plastic properties and composed of very fine particles; the clay mineral fraction of a soil is usually considered to be the portion consisting of particles finer than 2 μm; clay minerals are essentially hydrous aluminium silicates or occasionally hydrous magnesium silicates.

clinker. A partially fused product of a kiln, which is ground to make cement; also other vitrified or burnt material.

coefficient of thermal expansion. Change in linear dimension per unit length or change in volume per unit volume per degree of temperature change.

compressive strength. The measured maximum resistance of a concrete or mortar specimen to axial loading; expressed as force per unit cross-sectional area; or the specified resistance used in design calculations.

concrete. A composite material which consists essentially of a binding medium within which are embedded particles or fragments of aggregate; in Portland cement concrete the binder is a mixture of Portland cement and water.

continuous mixer. A mixer into which the ingredients of the mixture are fed without stopping, and from which the mixed product is discharged in a continuous stream.

coquina. A type of limestone formed of loosely or weakly cemented sea shells, found along present or former shorelines; used as a calcareous raw material in cement manufacture and other industrial operations.

curing. Maintenance of humidity and temperature of freshly placed concrete during some definite period following placing, casting or finishing to ensure satisfactory hydration of the cementitious materials and proper hardening of the concrete.

damp-proofing. Treatment of concrete or mortar to retard the passage or absorption of water or water vapour, by application of a suitable coating to exposed surfaces, by use of a suitable admixture or treated cement, or by use of preformed films such as polyethylene sheets.

deformation. A change in dimension or shape due to stress.

deterioration. Disintegration or chemical decomposition of a material during test or service exposure.

diatomaceous earth. A friable earthy material composed of nearly pure hydrous amorphous silica (opal) and consisting essentially of the frustules of microscopic plants called diatoms.

dicalcium silicate. A compound having the composition $2CaO \cdot SiO_2$, abbreviated C_2S, that occurs in Portland cement clinker.

dormant period. Interval during which the characteristics of a water–cement mix remain virtually unchanged.

early strength. Strength of concrete or mortar usually as developed during the first 72 h after placement.

endothermic reaction. A chemical reaction which occurs with the absorption of heat.

entrained air. Microscopic air bubbles intentionally incorporated in mortar or concrete during mixing, usually by use of a surface active agent; the bubbles are typically between 10 and 1000 μm in diameter and spherical or nearly so. (See **air entrainment**.)

epoxy resins. A class of organic chemical bonding systems used in preparing special coatings or adhesives for concrete or as binders in epoxy resin mortars and concretes.

ettringite. $3CaO \cdot Al_2O_3 \cdot 3CaSO_4 \cdot 30\text{–}32H_2O$, high sulphate calcium sulphoaluminate, also written as $Ca_6Al_2(SO_4)_3(OH)_{12} \cdot 26H_2O$, a mineral occurring in nature or formed by sulphate attack on mortar and concrete; the product of the principal expansion producing reaction in expansive cements.

evaporable water. Water in set cement paste present in capillaries or held by surface forces; measured as the water removable by drying under specified conditions.

exothermic reaction. A chemical reaction in which heat is evolved.

expanded shale (clay or slate). Lightweight vesicular aggregate obtained by firing suitable raw materials in a kiln or on a sintering grate under controlled conditions.

expansive cement (general). A cement which when mixed with water forms a paste that, after setting, increases in volume to a significantly greater degree than Portland cement paste; used to compensate for volume decrease due to shrinkage or to induce tensile stress in reinforcement (post-tensioning).

false set. The rapid development of rigidity in a freshly mixed Portland cement paste, mortar or concrete without the evolution of much heat, which rigidity can be dispelled and plasticity regained by further mixing without addition of water; premature stiffening, hesitation set, early stiffening and rubber set are terms referring to the same phenomenon, but false set is the preferred designation.

final set. A degree of stiffening of a mixture of cement and water greater than initial set, generally stated as an empirical value indicating the time in hours and minutes required for a cement paste to stiffen sufficiently to resist, to an established degree, the penetration of a weighted test needle; also applicable to concrete and mortar mixtures with use of suitable test procedures.

flash set. The rapid development of rigidity in a freshly mixed Portland cement paste, mortar or concrete, usually with the evolution of considerable heat, which rigidity cannot be dispelled nor plasticity regained by further mixing without addition of water; also referred to as quick set or grab set.

fly ash. The finely divided residue resulting from the combustion of ground or powdered coal and which is transported from the firebox through the boiler by flue gases.

free lime. Calcium oxide (CaO), as in clinker and cement, which has not combined with SiO_2, Al_2O_3 or Fe_2O_3 during the burning process, usually because of underburning, insufficient grinding of the raw mix or the presence of traces of inhibitors.

grout. A mixture of cementitious material and water, with or without aggregate, proportioned to produce a pourable consistency without segregation of the constituents; also mixtures of other compositions of similar consistency.

gypsum. A mineral having the composition calcium sulphate dihydrate ($CaSO_4 \cdot 2H_2O$).

haematite. A mineral, iron oxide (Fe_2O_3), used as an aggregate in high density concrete and in finely divided form as a red pigment in coloured concrete.

heat of hydration. Heat evolved by chemical reactions with water, such as that evolved during the setting and hardening of Portland cement. The difference between the heat of solution of dry cement and that of partially hydrated cement.

high early strength concrete. Concrete which, through the use of high early strength cement or admixtures, is capable of attaining a specified strength earlier than normal concrete.

hydrate. A chemical combination of water with another compound or an element.

hydrated lime. Calcium hydroxide, a dry powder obtained by treating quicklime with water.

hydration. Formation of a compound by the combining of water with some other substance; in concrete, the chemical reaction between hydraulic cement and water.

hydraulic cement. A cement that sets and hardens by chemical interaction with water and is capable of doing so under water.

hydraulic hydrated lime. The hydrated dry cementitious product obtained by calcining a limestone containing silica and alumina to a temperature short of incipient fusion, so as to form sufficient free calcium oxide to permit hydration and, at the same time, leaving sufficient calcium silicates unhydrated to give the dry powder its hydraulic properties.

initial set. A degree of stiffening of a mixture of cement and water less than final set, generally stated as an empirical value indicating the time in hours and minutes required for cement paste to stiffen sufficiently to resist, to an established degree, the penetration of a weighted test needle; also applicable to concrete or mortar with use of suitable test procedures.

kaolin. A rock, generally white, consisting primarily of clay minerals of the kaolinite group, composed principally of hydrous aluminium silicate, of low iron content; used as a raw material in the manufacture of white cement.

kaolinite. A common clay mineral having the general formula $Al_2(Si_2O_5)(OH)_4$, the primary constituent of kaolin.

latex. A water emulsion of a synthetic rubber or plastic obtained by polymerization and used especially in coatings and adhesives.

lime. Specifically, calcium oxide (CaO); also, loosely, a general term for the various chemical and physical forms of quicklime, hydrated lime and hydraulic hydrated lime.

matrix. In the case of mortar, the cement paste in which the fine aggregate particles are embedded; in the case of concrete, the mortar in which the coarse aggregate particles are embedded.

mixer. A machine used for blending the constituents of concrete, grout, mortar, cement paste or other mixtures.

monomer. An organic liquid, of relatively low molecular weight, that creates a solid polymer by reacting with itself or other compounds of low molecular weight, or both.

mortar. A mixture of cement paste and fine aggregate; in fresh concrete, the material occupying the interstices among particles of coarse aggregate; in masonry construction, mortar may contain masonry cement, or may contain hydraulic cement with lime (and possibly other admixtures), to afford greater plasticity and workability than are attainable with standard hydraulic cement mortar.

neat cement paste. A mixture of hydraulic cement and water, both before and after setting and hardening.

plaster of Paris. $CaSO_4 \cdot H_2O$, gypsum, from which three quarters of the chemically bound water has been driven off by heating; when wetted, it recombines with water and hardens quickly.

polyester. One of a large group of synthetic resins, mainly produced by reaction of dibasic acids with dihydroxy alcohols; commonly prepared for application by mixing with a vinyl group monomer and free radical catalysts at ambient temperatures and used as binders for resin mortars and concretes, fibre laminates (mainly glass), adhesives and the like. (See **polymer concrete**.)

polymer concrete. Concrete in which an organic polymer serves as the binder; also known as resin concrete; the term is sometimes erroneously employed to designate hydraulic cement mortars or concretes in which part or all of the mixing water is replaced by an aqueous dispersion of a thermoplastic copolymer.

polymerization. The reaction in which two or more molecules of the same substance combine to form a compound containing the same elements and in the same proportions, but of high molecular weight, from which the original substance can be generated, in some cases only with extreme difficulty.

porosity. The ratio, usually expressed as a percentage, of the volume of voids in a material to the total volume of the material, including the voids.

portlandite. A mineral, calcium hydroxide $(Ca(OH)_2)$; equivalent to a common product of hydration of Portland cement.

pozzolan. A siliceous or siliceous and aluminous material that in itself possesses little or no cementitious value but, in finely divided form and in the presence of moisture, will chemically react with calcium hydroxide at room temperature to form compounds with cementitious properties.

quicklime. Calcium oxide (CaO).

reactive silica material. Several types of materials which react at high temperatures with Portland cement or lime during autoclaving; includes pulverized silica, natural pozzolan and fly ash.

retardation. Reduction in the rate of hardening or setting, i.e. an increase in the time required to reach initial and final set or to develop early strength, of fresh concrete, mortar or grout.

retarder. An admixture which delays the setting of cement paste and hence of mixtures such as mortar or concrete containing cement.

Roman cement. A misnomer for a hydraulic cement made by calcining a natural mixture of calcium carbonate and clay, such as argillaceous limestone, to a temperature below that required to sinter the material but high enough to decarbonate the calcium carbonate, followed by grinding. So named because its brownish colour resembles ancient Roman cements produced by use of lime–pozzolan mixtures.

set. The condition reached by a cement paste, mortar or concrete when it has lost plasticity to an arbitrary degree, usually measured in terms of resistance to penetration or deformation; initial set refers to first stiffening, final set refers to attainment of significant rigidity; also strain remaining after removal of stress.

shale. A laminated and fissile sedimentary rock, the constituent particles of which are principally in clay and silt sizes; the laminations are bedding planes of rock.

shrinkage. Volume decrease caused by drying and chemical changes; a function of time but not of temperature or of stress due to external load.

silica flour. Very finely divided silica, a siliceous binder component which reacts with lime under autoclave curing conditions; prepared by grinding silica, such as quartz, to a fine powder; also known as silica powder.

silicone. A resin, characterized by water repellent properties, in which the main polymer chain consists of alternating silicon and oxygen atoms, with carbon-containing side groups; silicones may be used in caulking or coating compounds or as admixtures for concrete.

slump. A measure of consistency of freshly mixed concrete, mortar or stucco, equal to that subsidence measured to the nearest 6 mm of the moulded specimen immediately after removal of the slump cone.

sulphate attack. Chemical or physical reaction or both between sulphates, usually in soil or groundwater and concrete or mortar, primarily with calcium aluminate hydrates in the cement paste matrix, often causing deterioration.

sulphate resistance. Ability of concrete or mortar to withstand sulphate attack.

surface moisture. Free water retained on surfaces of aggregate particles and considered to be part of the mixing water in concrete, as distinguished from absorbed moisture.

talc. A mineral with a greasy or soapy feel, very soft, having the composition $Mg_3Si_4O_{10}(OH)_2$.

tensile strength. Maximum unit stress which a material is capable of resisting under axial tensile loading, based on the cross-sectional area of the specimen before loading.

tetracalcium aluminoferrite. A compound in the calcium aluminoferrite series, having the composition $4CaO \cdot Al_2O_3 \cdot Fe_2O_3$, abbreviated C_4AF, which is usually assumed to be the aluminoferrite present when compound calculations are made from the results of chemical analysis of Portland cement.

tobermorite. A mineral having the approximate formula $Ca_4(Si_6O_{18} \cdot H_2) \cdot Ca \cdot 4H_2O$ identified approximately with the artificial product tobermorite (G) of Brunauer, a hydrated calcium silicate having a CaO/SiO_2 ratio in the range 1.39–1.75, and forming minute layered crystals that constitute the principal cementing medium in Portland cement concrete; a mineral with 5 mol of lime to 6 mol of silica, usually occurring in plate-like crystals, which is easily synthesized at high steam pressures; the binder in several properly autoclaved products.

tobermorite gel. The binder of concrete cured moist or in atmospheric pressure steam, a lime rich gel-like solid containing 1.5–2.0 mol of lime per mole of silica.

trass. A natural pozzolan of volcanic origin.

tricalcium aluminate. A compound having the composition $3CaO \cdot Al_2O_3$, abbreviated C_3A.

tricalcium silicate. A compound having the composition $3CaO \cdot SiO_2$, abbreviated C_3S, an impure form of which (alite) is a main constituent of Portland cement.

vermiculite. A group name for certain platy minerals, hydrous silicates of aluminium, magnesium and iron; characterized by marked exfoliation on heating; also a constituent of clays.

water/cement ratio. The ratio of the amount of water, exclusive only of that absorbed by the aggregates, to the amount of cement in a concrete or mortar mixture; preferably stated as a decimal by weight.

water reducing agent. A material which either increases slump of freshly mixed mortar or concrete without increasing water content or maintains workability with a reduced amount of water, the effect being due to factors other than air entrainment.

water repellent cement. A hydraulic cement having a water repellent agent added during the process of manufacture, with the intention of conferring resistance to the absorption of water by the concrete or mortar.

wetting agent. A substance capable of lowering the surface tension of liquids, facilitating the wetting of solid surfaces and permitting the penetration of liquids into the capillaries.

workability. That property of freshly mixed concrete or mortar which determines the ease and homogeneity with which it can be mixed, placed, compacted and finished.

CONTRIBUTORS TO DRAFTING AND REVIEW

Bailey, G.	United Kingdom Atomic Energy Authority, United Kingdom
Bansal, N.K.	Bhabha Atomic Research Centre, India
Colombo, P.	Brookhaven National Laboratory, United States of America
De Angelis, G.	ENEA-CRE, Italy
Della Casa, A.	Société coopérative nationale pour l'entreposage de déchets radioactifs, Switzerland
Efremenkov, V.M.	International Atomic Energy Agency
Eriksson, A.	Swedish State Power Board, Sweden
Haighton, A.P.	International Union of Producers and Distributors of Electrical Energy
Jaouen, C.	Société générale pour les techniques nouvelles, France
Jia, Ruihe	Lanzhou Uranium Factory, China
Köster, R.	Kernforschungszentrum Karlsruhe, Federal Republic of Germany
Lee, D.J.	United Kingdom Atomic Energy Authority, United Kingdom
Michetti, F.	ENEA-DISP, Italy
Murillo, R.	Dirección de Technología Nuclear, Spain
Pulkkinen, R.	Oskarshamnsverkets Kraftgrupp AB, Sweden
Schweingruber, M.	Bundesamt für Energiewirtschaft, Switzerland
Simon, R.	Commission of the European Communities
Swennen, R.	ONDRAF/NIRAS, Belgium
Torok, J.	Atomic Energy of Canada Ltd, Canada
Vejmelka, P.	Kernforschungszentrum Karlsruhe, Federal Republic of Germany

Consultants Meetings

Vienna, Austria: 27–31 October 1986, 25–29 May 1987, 13–17 June 1988
Karlsruhe, Federal Republic of Germany: 14–18 September 1987

Technical Committee Meeting

Vienna, Austria: 18–22 May 1987

HOW TO ORDER IAEA PUBLICATIONS

 An exclusive sales agent for IAEA publications, to whom all orders and inquiries should be addressed, has been appointed for the following countries:

CANADA
UNITED STATES OF AMERICA   UNIPUB, 4611-F Assembly Drive, Lanham, MD 20706-4391, USA

 In the following countries IAEA publications may be purchased from the sales agents or booksellers listed or through major local booksellers. Payment can be made in local currency or with UNESCO coupons.

ARGENTINA	Comisión Nacional de Energía Atómica, Avenida del Libertador 8250, RA-1429 Buenos Aires
AUSTRALIA	Hunter Publications, 58 A Gipps Street, Collingwood, Victoria 3066
BELGIUM	Service Courrier UNESCO, 202, Avenue du Roi, B-1060 Brussels
CHILE	Comisión Chilena de Energía Nuclear, Venta de Publicaciones, Amunategui 95, Casilla 188-D, Santiago
CHINA	IAEA Publications in Chinese: China Nuclear Energy Industry Corporation, Translation Section, P.O. Box 2103, Beijing
	IAEA Publications other than in Chinese: China National Publications Import & Export Corporation, Deutsche Abteilung, P.O. Box 88, Beijing
CZECHOSLOVAKIA	S N T L, Spálená 51, CS-113 02 Prague 1
	Alfa, Publishers, Hurbanovo námestie 3, CS-815 89 Bratislava
FRANCE	Office International de Documentation et Librairie, 48, rue Gay-Lussac, F-75240 Paris Cedex 05
HUNGARY	Kultura, Hungarian Foreign Trading Company, P.O. Box 149, H-1389 Budapest 62
INDIA	Oxford Book and Stationery Co., 17, Park Street, Calcutta-700 016
	Oxford Book and Stationery Co., Scindia House, New Delhi-110 001
ISRAEL	YOZMOT Literature Ltd., P.O. Box 56055, IL-61560 Tel Aviv
ITALY	Libreria Scientifica Dott. Lucio di Biasio "AEIOU", Via Coronelli 6, I-20146 Milan
JAPAN	Maruzen Company, Ltd, P.O. Box 5050, 100-31 Tokyo International
PAKISTAN	Mirza Book Agency, 65, Shahrah Quaid-e-Azam, P.O. Box 729, Lahore 3
POLAND	Ars Polona, Foreign Trade Enterprise, Krakowskie Przedmieście 7, PL-00-068 Warsaw
ROMANIA	Ilexim, P.O. Box 136-137, Bucharest
RUSSIAN FEDERATION	Mezhdunarodnaya Kniga, Sovinkniga-EA, Dimitrova 39, SU-113 095 Moscow
SOUTH AFRICA	Van Schaik Bookstore (Pty) Ltd, P.O. Box 724, Pretoria 0001
SPAIN	Díaz de Santos, Lagasca 95, E-28006 Madrid
	Díaz de Santos, Balmes 417, E-08022 Barcelona
SWEDEN	AB Fritzes Kungl. Hovbokhandel, Fredsgatan 2, P.O. Box 16356, S-103 27 Stockholm
UNITED KINGDOM	HMSO, Publications Centre, Agency Section, 51 Nine Elms Lane, London SW8 5DR
YUGOSLAVIA	Jugoslovenska Knjiga, Terazije 27, P.O. Box 36, YU-11001 Belgrade

 Orders from countries where sales agents have not yet been appointed and requests for information should be addressed directly to:

 Division of Publications
International Atomic Energy Agency
Wagramerstrasse 5, P.O. Box 100, A-1400 Vienna, Austria